Top 20 Essential Skills for ArcGIS® Pro

TOP 20
ESSENTIAL SKILLS FOR
ARCGIS® PRO

Bonnie Shrewsbury, GISP
Barry Waite

Esri Press
REDLANDS | CALIFORNIA

Esri Press, 380 New York Street, Redlands, California 92373-8100
Copyright © 2023 Esri
All rights reserved.
Printed in the United States of America.
27 26 25 24 23   2 3 4 5 6 7 8 9 10

ISBN: 9781589487505
Library of Congress Control Number: 2023933567

The information contained in this document is the exclusive property of Esri or its licensors. This work is protected under United States copyright law and other international copyright treaties and conventions. No part of this work may be reproduced or transmitted in any form or by any means, electronic or mechanical, including photocopying and recording, or by any information storage or retrieval system, except as expressly permitted in writing by Esri. All requests should be sent to Attention: Director, Contracts and Legal Department, Esri, 380 New York Street, Redlands, California 92373-8100, USA.

The information contained in this document is subject to change without notice.

US Government Restricted/Limited Rights: Any software, documentation, and/or data delivered hereunder is subject to the terms of the License Agreement. The commercial license rights in the License Agreement strictly govern Licensee's use, reproduction, or disclosure of the software, data, and documentation. In no event shall the US Government acquire greater than RESTRICTED/LIMITED RIGHTS. At a minimum, use, duplication, or disclosure by the US Government is subject to restrictions as set forth in FAR §52.227-14 Alternates I, II, and III (DEC 2007); FAR §52.227-19(b) (DEC 2007) and/or FAR §12.211/12.212 (Commercial Technical Data/Computer Software); and DFARS §252.227-7015 (DEC 2011) (Technical Data–Commercial Items) and/or DFARS §227.7202 (Commercial Computer Software and Commercial Computer Software Documentation), as applicable. Contractor/Manufacturer is Esri, 380 New York Street, Redlands, California 92373-8100, USA.

Esri products or services referenced in this publication are trademarks, service marks, or registered marks of Esri in the United States, the European Community, or certain other jurisdictions. To learn more about Esri marks, go to: links.esri.com/EsriProductNamingGuide. Other companies and products or services mentioned herein may be trademarks, service marks, or registered marks of their respective mark owners.

For purchasing and distribution options (both domestic and international), please visit esripress.esri.com.

CONTENTS

Acknowledgments ix
How to use this book xi
Introduction xv

1 Exploring the ArcGIS Pro interface — 1
  Download census files — 2
  Add data to ArcGIS Pro — 5
  Explore tools and functionality — 8
  Explore the Contents pane — 10
  View tabular data — 11
  Modify shapefiles — 12

2 Creating reference maps and layouts — 17
  Create a reference map — 18
  Create a layout — 25

3 Preparing your data — 34
  Download census files — 35
  Prepare the table for ArcGIS Pro — 38

4 Joining tables to GIS data — 42
  Join a stand-alone table to a shapefile — 43
  Create a shapefile — 48

5 Creating thematic maps — 50
  Create a thematic map by symbolizing layers — 51
  Refine the ways you represent your data — 52

6 Geocoding — 63
  What's an address locator? — 64
  Use an address locator to geocode data — 64

7 Creating categorical maps — 71
  Examine the layer attributes — 72
  Display only selected types of facilities — 74
  Group multiple categories into a combined category — 76
  Create a layout and add elements — 78

8 Working with data tables — 80

- Add a shapefile and open the attribute table — 81
- Edit data in the attribute table — 81
- Edit data outside the attribute table on a polygon-by-polygon basis — 83
- Add a field to the attribute table — 84
- Calculate values — 86
- Change the number formatting — 88
- Delete fields — 89
- Export the altered shapefile — 90
- Work with multiple attribute tables — 91

9 Enriching your data — 93

- Add and symbolize data — 94
- Enrich the tracts — 94
- Review the table — 98
- Symbolize the layer — 98

10 Mapping x,y coordinate data — 102

- Add latitude and longitude data to ArcGIS Pro — 103
- Convert coordinates into points on your map — 103
- Add data from ArcGIS Living Atlas to your map — 105

11 Editing feature data — 109

- Add a shapefile — 110
- Edit a polygon's shape — 110
- Move a polygon — 112
- Split a polygon — 113
- Move a split polygon — 115
- Rotate a polygon — 116
- Scale a polygon — 117

12 Performing data queries — 120

- Add a shapefile — 121
- Write an attribute query — 121
- Examine the selection totals — 123
- Create a shapefile of the selected features — 124
- Clear the query — 125

13 Performing location queries — 127

- Add shapefiles — 128
- Write a location query — 129
- Export a table for the selected features — 130

14 Using geoprocessing tools — 133
- Pairwise Buffer tool — 134
- Merge tool — 141
- Append tool — 142
- Pairwise Clip tool — 144
- Pairwise Dissolve tool — 147

15 Creating geodatabases — 151
- Explore the Catalog pane — 152
- Create a file geodatabase — 152
- Import shapefiles into a geodatabase — 154
- Import a table into a geodatabase — 155
- Load an aerial photo into a geodatabase — 156

16 Joining features — 160
- Add data — 161
- Select storm events for only your state — 161
- Export the selected points — 162
- Perform a spatial join — 163
- Perform a second spatial join — 165
- Map the storm counts by county — 167

17 Working with imagery — 170
- Add image data — 171
- Georeference the image — 171

18 Using 3D data — 180
- Convert a 2D project into a 3D project — 181
- Improve the look of the 3D scene — 183
- Change building symbology — 184
- Extrude the BuildingFootprints layer — 185

19 Adding a table and chart to a layout — 189
- Insert a table frame — 190
- Modify the fields in the table — 191
- Sort the values of the Table Frame element — 193
- Create a chart — 194
- Insert a chart — 195

20 Sharing your work — 198
- Export a layout — 199
- Create a layer file — 200
- Add the layer file to your map — 201
- Import symbology from a layer file — 202
- Create a map package — 203
- Create a project package — 205

21 Publishing your work (bonus skill) — 209

- Share a web layer — 210
- Share a web map — 216
- Explore options for creating an app — 220
- Conclusion — 221

ACKNOWLEDGMENTS

Thank you to Gina Clemmer, author of *The GIS 20: Essential Skills for ArcMap*™ (Esri Press, 2017), which included the most valuable skills for GIS written in a way that was not only clear but engaging. Our students enjoyed the book so much that they kept it—a rare thing for college students!

Speaking of our students, we thank them for trying out our lessons and helping us make the final product all it could be. Our students inspire us and encourage us to do our best.

Thank you to Esri for always being there for us as GIS users, instructors, and now authors.

Last, thank you to our families and friends, who were so supportive during this process.

HOW TO USE THIS BOOK

About this book

Top 20 Essential Skills for ArcGIS® Pro has been tested for compatibility with ArcGIS Pro 3.1 software and is designed for students in a classroom or others who want to learn GIS on their own. No prior GIS knowledge or experience is needed.

Software requirements and licensing

To perform the exercises in this book, you'll need the following: ArcGIS Pro 3.1 installed on a computer that's running the Windows operating system, an internet connection, Microsoft Excel, and a web browser to access ArcGIS Online. Earlier software versions may not be fully compatible with this exercise data and may not operate as described in the exercises. Hardware requirements for ArcGIS Pro are available at links.esri.com/SysReqs.

Information on software trial options, as well as Personal Use and Student Use licensing, can be found at esri.com.

Use an existing license

If you have existing credentials, or can obtain credentials from your educational institution or organization, for accessing the required software listed in this section, you may use those credentials and proceed.

Use a 180-day software trial, download, and install ArcGIS Pro

1. Locate the software trial code that comes with participating books.

    - Hard-copy textbooks purchased in the Americas come with a code printed inside the back cover for a free trial.

- E-books purchased through the delivery platform VitalSource may come with free trials; participating e-books will be labeled *courseware*. With every e-book purchased, a code will be provided by VitalSource. Visit links.esri.com/FindTrialCode for help with locating this code.

2. Use your trial code to sign up at links.esri.com/TrialCode.

3. Once you have signed up for your software trial, license ArcGIS Pro.

    a. Sign in at arcgis.com.

    b. On the home page, click Organization, and click the Licenses tab.

    c. Locate the license for ArcGIS Pro Advanced, and click Manage.

    d. Next to your username, turn on the license, and close the Settings pane.

4. Download ArcGIS Pro.

    a. Sign in at arcgis.com.

    b. On the top toolbar, click your profile icon in the upper right, and click Trial Download.

    c. On the ArcGIS Free Trial page, click Download ArcGIS Pro.

5. When the download of ArcGIS Pro is complete, open the file and install it. If you encounter problems during installation, visit the guide at links.esri.com/ProInstallHelp.

Installing the exercise data

The exercise data for this book is available at links.esri.com/Pro20Data. Download the exercise data, and extract it to your local (C:) drive.

Downloadable data that accompanies this book is covered by a license agreement that stipulates the terms of use. Review this agreement at links.esri.com/LicenseAgreement.

Resources, feedback, and updates

The ArcGIS Pro Help documentation provides comprehensive descriptions of software concepts and tools at links.esri.com/Help.

Feedback, updates, and collaboration are available at Esri® Community, the global community of Esri users. Post any questions about this book at links.esri.com/EsriPressCommunity.

Visit the book's web page at links.esri.com/Pro20.

Tips for completing the chapters

Start at the beginning

This book is designed for chronological progression—earlier chapters have more explicit instruction than later chapters. Exercises within chapters build on one another, so it's advisable to perform all the exercises within a chapter in order.

Just try

Everything is spelled out in each chapter. If you get lost, back up and see whether you've missed something. We've helped hundreds of students get through problems this way, and you'll get through it, too!

Don't be just a button clicker

Some functions take a few seconds or even minutes to complete. You may be using data on your computer or on a server on another continent, which can slow things down. Don't start clicking away randomly, hoping to speed things along. Just take a second and breathe!

Don't panic if something goes wrong

All GIS software is complex behind the scenes. You're taking information from a round Earth, displaying it in a flattened visual format, and performing analyses on that flattened version. The software does all the hard stuff for you, but no software can know every problem that may occur, and glitches happen. Don't take it personally. The great news is that you can usually find at least three ways to do anything in ArcGIS Pro, so you can try an alternative if one doesn't work for you.

Don't rush

There are no points given for finishing quickly. You aren't in a race. If you're like us, you're task oriented and just want to finish. Remember that finishing each lesson is not your goal. Learning the skills is your goal. Slow. Down.

Read everything

Part of not rushing is to make sure you read all the text. Don't scan for instructions only. You need to understand what you're doing and why. The explanations we provide help you understand not only how to but why you'll perform these steps. If you perform some steps and then wonder, "What did I just do?" you'll need to reread the section.

Getting help

ArcGIS Pro Help

ArcGIS Pro includes both an online and an offline Help system that's handy for questions about how a function operates. If you have a question, it's a great place to start. Visit the online Help at links.esri.com/Help. To access the offline Help, click the View Help button at the top of the application window.

Researching error messages

If you see an error message that has you scratching your head, just search for it in your browser. For example, let's say you're running Microsoft Excel and see this common error message: "Required Microsoft driver is not installed." In your browser, search for "ArcGIS Pro Required Microsoft driver is not installed." Combine the product name with the keywords. The relevant information will appear in your search results.

YouTube and user communities

GIS users have created a huge array of YouTube videos about issues ranging from the simplest tasks to the most complicated. In general, you'll find the GIS community to be a great group of people willing to help any way they can. Check out Esri Community, a global community of Esri users, at community.esri.com.

INTRODUCTION

Geographic information systems (GIS) are valuable in almost any field. There may be a field in which geography doesn't matter, but we can't think of any. In business, suppliers and customers need locations. In government, whatever doesn't have an address can be located with a GPS point. In environmental studies, there's a geographic component to waterways, landforms, animals, plants, soil types, air quality, and so much more. GIS is an invaluable tool in all these fields, and it's something you can learn to use yourself.

In the last 20 years, we've worked with a variety of students, most of whom had no GIS knowledge before taking our classes. Some of them considered themselves very much nontechnical and were fearful of taking on what they saw as a huge challenge. Afterward, they were amazed at how much they could do on their own with GIS. Yes, users can perform complicated and challenging tasks using GIS, but you can also perform many tasks with just a bit of learning and practice. That's where we can help.

As you've noticed, *Top 20 Essential Skills for ArcGIS Pro* isn't a large book. Each chapter introduces you to a useful, accessible skill, including how to create reference maps and layouts, edit feature data, and perform queries, and lets you practice each skill with hands-on exercises and data. But why 21 chapters for 20 skills? We've given you a bonus skill—publishing your work in ArcGIS Pro. Each chapter also includes a user story that lets you explore how these skills have been used in the real world to answer important questions and solve problems. When you've completed the book, you can use the table of contents to locate specific tasks you want to do again for practice.

After completing this book, you may want to continue your GIS journey with any number of lengthier or more advanced books (including several from Esri Press). We don't expect you to become a GIS expert in these short chapters, but if you take this journey with us, you'll be able to use many important GIS tools to work with geographic data.

CHAPTER 1

Exploring the ArcGIS Pro interface

Before you get started, you'll need to install the software and download the exercise data, as described in the "How to Use This Book" section. If you haven't read that section, read it now—you'll need that information.

Let's start with downloading some shapefile data. So, what's a shapefile? It's a file that stores the location, shape, and tabular attributes of geographic features. The geographic features can be points, lines, or polygons. Although more sophisticated data storage types are available, shapefiles are a common building block for creating maps and performing analyses in a geographic information system (GIS).

In this chapter, you'll download shapefiles from the US Census Bureau and learn to use some of the key tools in ArcGIS® Pro, a GIS desktop program that uses a ribbon, similar to Microsoft Word. The ribbon contains tabs, groups, and tools that are organized to help you perform your work. This style of interface should already be familiar to you.

Download census files

Access the US Census Bureau archive

1. In a browser, browse to **links.esri.com/Pro20_Census** to visit the US Census Bureau data download location.

2. Click the tab for 2022.

3. At the bottom, under Download, click FTP Archive.

   This link takes you to the bureau's FTP archive.

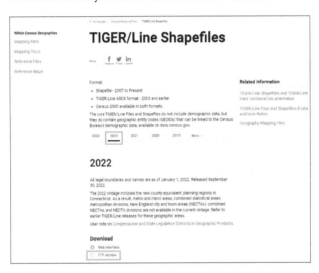

Download county, state, and place shapefiles

1. In the FTP archive, click the County folder to open it.

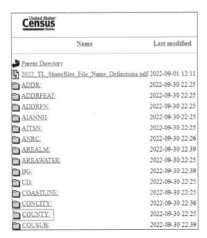

2. Click tl_2022_us_county.zip to download the zip file to your Downloads folder.

   Downloading the zip file may take several minutes.

   You'll extract this zip file after you've finished downloading it.

3. Let the file continue to download, and click the browser's back arrow to select additional files.

4. Scroll down in the folder directory, and click the State folder.

5. Click tl_2022_us_state.zip to download the zip file to your Downloads folder.

6. While the download progresses, click the browser's back arrow.

7. Scroll down in the folder directory, and click the Place folder.

   The US Census Bureau uses the name *Place* to refer to cities. This folder includes all US cities, as well as census designated places (CDPs), which are not cities per se but generally recognized areas. The important takeaway here is that the Place folder contains the city files.

   The Place folder lists many zip files because the list includes files for all states and territories. This book concentrates on the authors' home state of California, which is state 06.

   If you use another state's place file, your project won't exactly match the examples or instructions used throughout the rest of the book. For example, instructions that say to open the CaliforniaCounties file won't match what you're seeing on your screen. If you're comfortable substituting the name and data for your state, feel free to do so. You can open a new tab in your browser and search **[YourState] FIPS code**. *This code is the number you'll need instead of 06 when downloading the place shapefile.*

8. Click tl_2022_06_place.zip to download the zip file to your Downloads folder.

FIPS codes

FIPS stands for Federal Information Processing Standard, which is a unique identifier (ID) for geographic areas. States have two-digit codes, and counties have three-digit codes. A state code plus a county code is a five-digit unique ID for every county in the United States.

> tl_2022_06_place.zip 2022-10-31 19:42 9.3M

You now have all the GIS census data you need.

9. Open File Explorer, and browse to your Downloads folder.
10. Right-click one of the zip files, click Extract All, and save the files to C:\GIS20\Chapter01. Repeat this step for the remaining shapefiles.

What files make up a shapefile?

After unzipping the US Census Bureau files, you'll have a file with the .shp file name extension. You'll also have a few other files with different file name extensions. A shapefile is made up of all these files, not just the .shp file. Some of them are mandatory and some are optional.

These are the mandatory files:

- .shp = the geometry of the features in the layer.
- .dbf = database file that stores the tabular data for the features.
- .shx = index file that ties the .shp and .dbf files together.

These are the optional files:

- .prj = projection file that lists the map projection being used.
- .shp.xml contains metadata about the shapefile.
- .sbn and .sbx = spatial index files.
- .cpg describes the encoding applied to create the shapefile.

Other file types may also be associated with a shapefile, but they are beyond the scope of this book.

Add data to ArcGIS Pro
Start ArcGIS Pro

1. Click the Windows button on the taskbar to open the Start menu. In the alphabetized list of programs, click ArcGIS to expand it. In the folder, click ArcGIS Pro. Ignore all the other programs—ArcGIS Pro is the mapping program.

   When ArcGIS Pro starts, you have several choices for the type of project you want to start. An ArcGIS Pro project is a collection of related maps, layouts, and resources (such as system folders and databases). Project files have the .aprx file name extension. Projects are stored in their own folders with their own databases (known as geodatabases). You're going to start with a map this time.

2. Under New Project, click Map.

3. For Name, type **MyState**.
4. For Location, browse to your C:\GIS20\Chapter01 folder.

   You are telling ArcGIS Pro which folder to save your ArcGIS Pro project in. You want *Chapter01* to show in the Name box.

5. Click OK to create the new project.

   At the top of the project window, your project name is shown.

Add shapefiles to an ArcGIS Pro project

ArcGIS Pro uses a ribbon interface that groups commands into categories. It can take a bit of getting used to, but it's handy to have.

1. On the Map tab, in the Layer group, click Add Data to begin adding the shapefiles to your project.

   If you click the top part of the Add Data button, the Add Data dialog box will open. If you click the down arrow, you'll see a list of data types from which to choose. Most of the time, you can click the button and not the down arrow, because you'll generally be adding data from your computer.

You will use this tool constantly, so it's a good idea to get to know it.

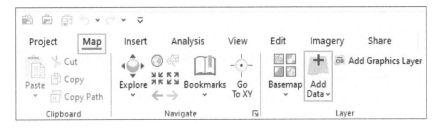

The challenge is to browse to the folder where you unzipped the three shapefiles.

2. In the left pane, scroll down and click This PC. In the right pane, double-click Local Disk (C:), double-click the GIS20 folder, and double-click the Chapter01 folder to locate your shapefiles.

    If you don't see the shapefiles (they have the .shp file name extension), you haven't properly unzipped the files, or they aren't in that folder. If you've unzipped them into individual folders, that's OK—you'll need to look in each folder and add the files individually. If you still don't see them, redo the earlier instructions, unzipping and adding them again. If you think the problem is something else, we assure you it isn't.

3. Click tl_2022_06_place.shp, and then press and hold the Ctrl key while clicking both the county file and the state file. Click OK to add all three shapefiles.

    ArcGIS Pro randomly assigns the colors. You can change them later.

    The map expands to show the whole world. Not so helpful, but we can fix that.

4. In the Contents pane, right-click tl_2022_06_place.shp, and choose Zoom to Layer.

    This layer has data for only one state, so the map zooms to that state's extent.

Chapter 1: Exploring the ArcGIS Pro interface

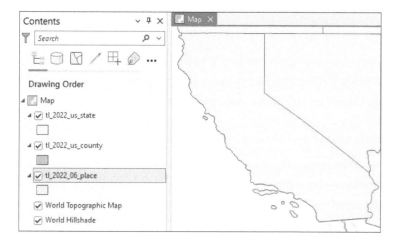

Reorder layers

The shapefiles will be added to the map with the state layer at the top of the Contents pane. The layers in the Contents pane draw in a specific order: from the bottom up. Because the state layer draws last, it covers the county and place layers.

1. At the top of the Contents pane, ensure that the List by Drawing Order tool is selected.

2. In the Contents pane, click tl_2022_us_state and drag it down so it's directly above the World Topographic Map. Drag tl_2022_06_place to the top of the list.

   Now your map looks better!

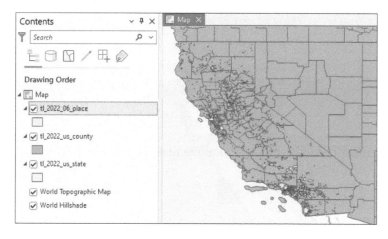

Explore tools and functionality

Because ArcGIS Pro provides hundreds of tools, it's essential to identify the most important ones. The tools you'll use in nearly every mapping session are featured in this section. Try to become familiar with these tools and what they can do.

Use the Add Data tool

You can add other shapefiles, layer files, or data tables to your project the same way you added the census shapefiles—by using the Add Data tool. This tool will be one of your best friends as you progress.

To add data, you click the Add Data tool on the Map tab, in the Layer group. If you don't see the Add Data tool, make sure you've selected the Map tab first.

Use the scroll wheel to zoom in and out

You can navigate the map by using the scroll wheel on your mouse or pinching your laptop touch pad to zoom in and out. Here's a helpful explanation of how the mouse interacts with the map.

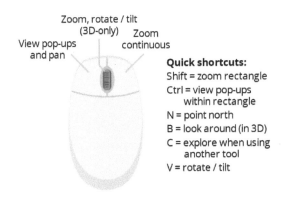

1. Use your mouse functionality to practice zooming in and out.

Use the Fixed Zoom In and Fixed Zoom Out tools

The Fixed Zoom In and Fixed Zoom Out tools are found on the Map tab, in the Navigate group. They're used to zoom in or out a fixed amount at a time.

1. Use the Fixed Zoom In and Fixed Zoom Out tools to practice.

Use the Explore tool

To fully navigate the map, you'll need to use the Explore tool, which allows you to pan and zoom.

1. On the Map tab, in the Navigate group, click the Explore tool.

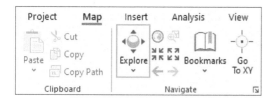

   A hand with a pointing finger indicates that the tool is now active.

2. Use the Explore tool to drag and pan the map to practice.

   The Explore tool is the default tool for navigating maps in ArcGIS Pro. It can also be used to identify features on the map. Every map layer in ArcGIS contains both the features that draw on the map and a table of attribute information that accompanies the features. The attribute table contains information such as name, address, type, and so on. Identifying a feature means seeing that feature's attribute information in a pop-up window.

3. With the Explore tool active (check that the cursor still looks like a hand with a pointing finger), click a feature of your choice to see a pop-up window showing that feature's tabular attribute information.

Use the Full Extent tool

The Full Extent tool zooms to the full area of all the layers in your map. It's a good way to center your map on your screen.

> *When you're using a basemap, the Full Extent tool zooms to the whole world, which is the full extent of the basemap. Not as helpful. We'll talk about basemaps in another chapter.*

1. Click the Full Extent tool to reposition your map.

Use the Previous Extent and Next Extent tools

Sometimes you'll zoom to a map location you didn't want. It happens to everyone. You can easily use the Previous Extent and Next Extent tools to return or advance to preferred extents. The blue arrows on the ribbon indicate existing previous or next extents.

1. Click the Previous Extent and Next Extent tools to see how they work.

Explore the Contents pane

The Contents pane, located on the left side of the ArcGIS Pro window, allows you to work with files in your map. The Contents pane is where your data layers are listed. The top of the Contents pane features a search box and a number of tabs (more or fewer depending on your license level) that control layer management.

1. Examine the tabs in the Contents pane: *from left*, List by Drawing Order, List by Data Source, List by Selection, List by Editing, List by Snapping, and List by Labeling. Point to any of these tabs to see a description of the function.

2. Click the List by Drawing Order tab (the first icon from the left).

   List by Drawing Order is the default active tab when you open a new project.

3. Use the scroll wheel to zoom in to your state.

4. Check the box to the left of a layer to turn it on or off.

   When a layer is on, it appears in the map view. When a layer is off, it disappears in the map view (but is not deleted).

   As you learned earlier, you can move layers up or down in the Contents pane by clicking a layer once and dragging it to the desired position. When you do so, the map changes as you reposition the layers in different drawing orders. When the place layer is at the top, the cities in your state appear in the map. When the county layer is at the top, it draws last, and because it has a solid fill color, it covers the place layer so that the cities disappear in the map. Before you move on, make sure the place layer is at the top.

   The layer names are used in the legend. You can make them easier to understand or remember by renaming them.

5. For each of the layers, click the layer name twice to activate the text (two deliberate separate clicks, not a double-click). Type **Cities**, **Counties**, and **States** over the existing layer names to make them easier to understand.

   > *When you rename a layer in the Contents pane, the layer is not renamed in the underlying data but only in your map.*

View tabular data

Shapefiles contain many things, including the geometry of the layer and the data for each feature. So far, you've examined the map layer part of the shapefile and used the Explore tool to peek at the underlying data. Another way to view a shapefile's data is to view the whole data table, or attribute table, at once.

1. In the Contents pane, right-click Counties (the new, easily understandable layer name), and click Attribute Table from the list.

2. Use the scroll bars to examine the contents of the data table.

   You'll notice a few thousand counties in the attribute table. The table contains no demographic data, only FIPS codes, county names, and other miscellaneous codes and information that the US Census Bureau included when it created the shapefile.

3. Right-click any of the column headings, and click Sort Ascending.

   This command sorts the entries alphabetically (if the column contains text) or from the smallest to largest value (if the column contains numbers). Sorting is useful when you want to locate entries in a large group.

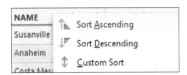

4. Close the attribute table by clicking the *X* to the right of the table name in the upper-left corner.

Modify shapefiles

Shapefiles downloaded from the US Census Bureau, or anywhere else, may need to be modified to suit your purposes. A great example: when you downloaded the county shapefile, you had no option to download a county file for just one state. The only file offered contained all the counties for the entire United States.

For the following exercises, it would be better to have a shapefile of only those counties in your state (like the place shapefile). How do you do that? It's easy to modify shapefiles. In the next few steps, you'll isolate your state's counties and create a new shapefile that contains only those counties.

1. In the Contents pane, right-click the Cities layer, and click Attribute Table.

2. Find the StateFP column (third column from the left), and note your state's two-digit FIPS code. (Reminder that for California, the FIPS code is 06.)

3. Close the attribute table.

4. In the Contents pane, right-click the Counties layer, and click Attribute Table.

   Opening the attribute table for a layer is a common task you'll be doing throughout this book. Now you know how to open the attribute table—another essential skill!

5. Right-click the StateFP column heading, and click Sort Ascending to organize the table by state.

6. Scroll down to find the first record in the StateFP column that matches your state's FIPS code.

7. After you find the first record for your state, click the first gray cell at the left of that row.

   Clicking the cell highlights that record in cyan (bright greenish blue).

FID	Shape	STATEFP ▲	COUNTYFP	COUNTYNS	GEOID	NAME
186	3105	Polygon	05	111	00069174	05111	Poinsett
187	3222	Polygon	05	057	00066858	05057	Hempstead
188	8	Polygon	06	091	00277310	06091	Sierra
189	325	Polygon	06	067	00277298	06067	Sacramento
190	329	Polygon	06	083	00277306	06083	Santa Barbara

8. Scroll down through the table until you find the last record for your state, hold the Shift key, and click the first gray cell to the left of that row.

   You've now highlighted all the county records for your state. The counties are highlighted on the map, too. The attribute table and map features are linked. You may need to zoom in or pan the map to see your state more clearly.

Next, you're going to create a new shapefile for only those counties you selected.

9. In the Contents pane, right-click the Counties layer. Click Data > Export Features.

10. For Output Feature Class (the output location), click the Browse button (open folder icon), and browse to your C:\GIS20\Chapter01 folder.

11. For Name, type your state name followed by **Counties**. For example, type **CaliforniaCounties** (don't use spaces in shapefile names!). The .shp file name extension is added for you. Click Save.

12. Click OK to export the file.

    The new shapefile is automatically added to your project at the top of the Contents pane. Now you can organize things. The original layer with all the US counties is no longer needed, so you can remove it.

13. If you have an attribute table that's open, close it.

14. In the Contents pane, right-click Counties (this is your original county file), and click Remove.

15. Drag the Cities layer back to the top of the Contents pane.

    You've done good work! Let's make sure to save it.

16. At the top of the project window, click the Save button (third button from the left).

Sharing an ArcGIS Pro project

Saving and sharing a project is more complicated than it seems. An ArcGIS Pro project (which uses the .aprx file name extension) contains your maps, layouts, and links to all the data sources that make up the project. If you send the file to coworkers, they won't be able to open it without also having all the data sources that make up the project. We'll talk later about options for packaging your project and data layers for sharing.

Congratulations! You made a map! You downloaded shapefile data from the US Census Bureau, added that data to a project in ArcGIS Pro, learned tips for navigating the map, exported the part of the data you needed, and saved your project to build on later. Update your résumé to show that you have experience with ArcGIS. OK, that may be a little premature, but you've taken a big step in the right direction. Onward!

17. Click the *X* in the top-right corner to close ArcGIS Pro.

USER STORY

The variety of census data

This first user story is not the story of a particular GIS user (in later chapters, you'll get real user stories). Instead, let's look briefly at the variety of data that's available from the US Census Bureau. This list contains some examples of data types:

- Population
- Age
- Gender
- Race
- Employment status
- Education level
- Income
- Homeowner or renter status
- Country of origin
- Language spoken
- Vehicle ownership
- Commute mode

And so many more! These data types can be combined in all sorts of ways to answer questions about people in a particular location. How old are the people in primarily renter-occupied areas versus in primarily homeowner-occupied areas? Where are people who have limited vehicle ownership or low income or who speak another language—and may need transit information in that language? Where are people working in agriculture who come from another country and have children who may need reading help?

From a business standpoint, where are the potential customers for a day-care center? If I put a store in a particular area, do I need to be concerned with transit access? If I offer a loan program for first-time home buyers, where is the greatest need?

Census data is useful in so many ways. Once you get the hang of it, don't be surprised if you come up with more questions—questions for which ArcGIS and census data can provide the answer!

CHAPTER 2

Creating reference maps and layouts

Reference maps show the location of geographic features. They display the boundaries, names, and unique identifiers of standard geographic areas, as well as major cultural and physical features, such as roads, railroads, coastlines, rivers, and lakes. Examples of reference maps include general atlases and road maps. As perhaps one of the most basic types of maps, reference maps are a good place to start. You'll make more sophisticated types of maps in later chapters.

Congratulations on getting your data and mapping it in the last chapter! Now you'll learn how to turn that project into something for people to use. In this chapter, you'll make your first map layout, a reference map. You'll also learn how to add layout elements (such as a north arrow, scale bar, title, and legend) to your map in ArcGIS Pro. These layout elements are important because they give the audience context for understanding the message your map is intended to convey. How do you make your message clear? Let's do it.

Create a reference map

Open the project

1. Start ArcGIS Pro.

   > If you don't remember how, review the "Start ArcGIS Pro" section in chapter 1.

2. If prompted, sign in to ArcGIS Online with your account username and password.

3. To open your saved project from chapter 1, under Recent Projects, click MyState.

4. In the Contents pane, right-click CaliforniaCounties (or your state's counties layer), and from the options, click Zoom to Layer so the map zooms to your state.

Set a bookmark

You just zoomed to your state, so it's a good time to learn about a handy feature—the Bookmarks tool.

1. On the Map tab, in the Navigate group, click Bookmarks.

2. From the drop-down menu, click New Bookmark.

3. Enter a name and a description (optional), and click OK.

4. Click the Explore button, and pan the map away from your state.

5. Click the Bookmarks tool again, and click the bookmark you just created.

   Pretty cool, right? Bookmarks can help when you're working on something that requires you to move around a lot on the map. You can make as many bookmarks as you need. Just make sure to give them useful names so that you can keep track of what area each is bookmarking.

Change layer colors

You can change the fill and outline colors for each layer in ArcGIS Pro. These color values of a layer are known as symbology. To symbolize the CaliforniaCounties layer, follow these steps.

1. In the Contents pane, right-click CaliforniaCounties, and click Symbology from the list.

   The Symbology pane opens on the right side of the map.

2. Click the solid-color square symbol.

   > ArcGIS Pro assigns the default color randomly, so your layer may be a different color from what's shown in the image. That's okay.

   Tabs for Gallery and Properties appear in the Symbology pane.

3. Click the Properties tab. Under Appearance, click the Color square symbol to access the color palette, which provides choices for colors.

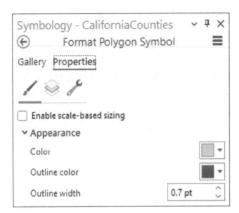

At least 120 different colors and shades are there for you to choose. You may also decide to use no color for a fill.

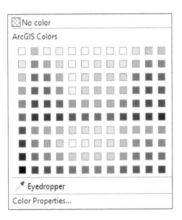

4. Click No Color to give the counties a hollow fill.

5. Click Apply at the bottom of the Symbology pane.

   A hollow fill allows your users to see features of other layers that lie below it in the Contents pane drawing order. You can also change the outline color and outline width of your features. Just remember to click Apply to see your changes.

6. In the Contents pane, uncheck the box for the states layer to turn it off.

   Now you can see the basemap under the counties.

7. Follow the same steps to change the cities layer symbology:

   a. In the Contents pane, right-click Cities, and click Symbology.

   b. Click the solid-color square symbol.

c. On the Properties tab, click the color square symbol, and choose a color.

d. Click the color square symbol for Outline Color, and click No Color. (Some layers are easier to see if they have no outline.)

e. Click Apply to see your changes to the cities layer.

8. Close the Symbology pane.

> **Hint:** Click the *X* in the upper-right corner.

This image illustrates the latest symbology changes to your map.

Change layer transparency

1. In the Contents pane, make sure the cities layer is turned on and selected.

2. On the Feature Layer tab, in the Effects group, for Transparency, type **50** or click the down arrow to set the value.

   The cities layer is now 50 percent transparent, allowing you to see through it.

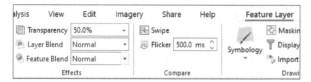

Transparency is helpful when you want to see features beneath a layer yet still be able to see the layer for reference. Here, you can see the county outlines beneath the cities layer because the cities are 50 percent transparent.

Label your features

1. Zoom to the San Francisco region.

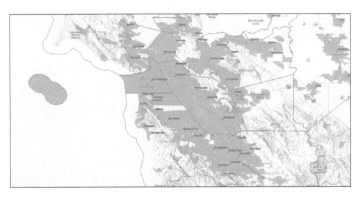

2. Zoom back to CaliforniaCounties.

3. In the Contents pane, right-click the cities layer. Click Label from the list (not Labeling Properties!) to turn on labeling.

It may take a while to draw the labels, but that's fine. You can monitor the spinning progress icon in the lower right of the map to see that the task is progressing. Don't click other buttons while your labels are being drawn.

As the process runs its course, the cities in the map will be labeled with their Name value from the attribute table. They're a bit of a mess, right? You can make the labels easier to read.

Change label settings

Depending on what you don't like about your labels, you can change several settings to improve them:

- Remove duplicate labels.
- Assign a buffer distance to the labels so they don't all draw on the map.
- Create a halo effect around the labels so they stand out better.

You'll do all of these next. (Yes, all three. You'll be glad you did!)

Remove duplicate labels

1. In the Contents pane, right-click Cities, and click Labeling Properties from the list to open the Properties pane.

   The Class tab opens by default (the default is indicated with a blue underline).

2. To the right of the Class and Symbol tabs, click Position.

3. Click the Conflict Resolution icon.

   Hint: Point to each icon to see a pop-up label, or ScreenTip, to find Conflict Resolution.

4. Expand Remove Duplicate Labels, and change the selection to Remove All.

   Alternatively, you can choose Remove Within Fixed Distance, but for now you'll use Remove All to remove the duplicate labels.

   Keep monitoring the spinning progress icon while you wait for the labels to be redrawn.

Assign a buffer distance

5. Under Conflict Resolution, expand Minimum Feature Size.

6. Change the specifications to **Area**, **2.0**, and **Millimeters**.

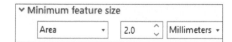

Setting these values tells ArcGIS Pro to draw labels only for cities that are more than two millimeters in area on the map. Feel free to change the values until the labels look good for your state.

Create a halo effect

7. To add a halo and make other changes to the text styling, click the Symbol tab.

8. Click Appearance to expand it, change Font Style to Bold, and click Appearance again to collapse it.

9. Expand Halo, click the Halo Symbol down arrow, and choose White Fill (the first choice under Polygon Symbols).

10. Click Apply at the bottom of the pane.

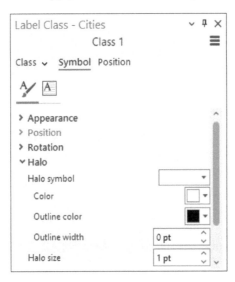

Again, it might look as if it didn't work, but check the spinning progress icon. Yep, it's still drawing!

11. When the label-drawing process finishes, click the X in the upper right of the Label Class pane to close it.

12. Zoom to the whole-state bookmark.

> **Hint:** On the Map tab, in the Navigate group, click Bookmarks, and click your bookmark.

This is a good time to save your project because you have made many changes. Get in the habit of saving whenever you've made changes. It'll save you time and grief!

13. At the upper left, click Save.

14. Examine the changes to your map.

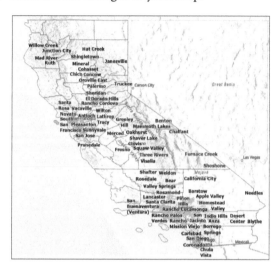

Create a layout

So far, you've been working with a map. That's where the data layers are added and symbolized to look just right. But when you're ready to share your map, you'll probably want a layout that showcases your map with some useful and appealing elements. In the layout, you can add a title, north arrow, scale bar, legend, and other elements. First, let's create an ArcGIS Pro layout so that we can examine the relationship between maps and layouts.

1. On the ArcGIS Pro ribbon, click the Insert tab.

2. Click New Layout.

   Your options include either a portrait or a landscape orientation in different sizes.

3. For this layout, choose a portrait orientation and an 8.5-by-11-inch page size.

   A new Layout tab is added next to your existing Map tab. You can tell that it's a layout because it looks like a piece of paper with rulers to the left and top.

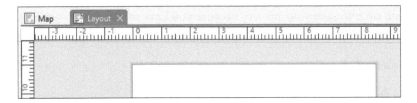

Add a map to the layout

But the piece of paper is blank. Where's your map? You can add it to the layout as a map frame.

1. On the Insert tab, in the Map Frames group, click the Map Frame button.

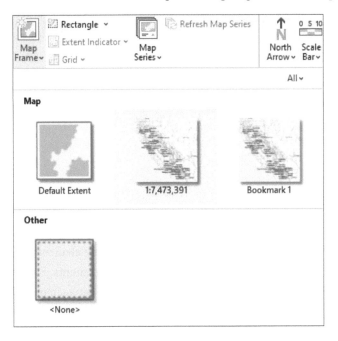

Under the Map section, you can find a Default Extent map and a thumbnail of your map with the map scale provided beneath it. There's even a choice to add your bookmark extent.

2. Click the thumbnail of your map.

The down arrow disappears and the cursor changes, but the map is still not in your layout. That's odd. Nope—ArcGIS Pro doesn't know where you want to put the map in your layout. Here's how to add it.

Chapter 2: Creating reference maps and layouts

3. Click somewhere in the upper left of the frame, and drag to draw a box on the layout to indicate where you want your map placed.

   The size and placement can be changed later, so don't worry about being precise.

4. Click back and forth on the Layout and Map tabs, toggling between the two so you familiarize yourself with what each looks like and how they are different. Examine how the Contents pane changes in each one.

Modify and insert layout elements

The layout is now listed in the Contents pane. Having it in the Contents pane gives you another place to access the elements and modify their properties.

Change the layout orientation

1. To change the orientation of the layout, in the Contents pane, under Drawing Order, right-click Layout, and click Properties.

   > **Hint:** You can also just double-click the word *Layout* to open the Layout Properties dialog box directly. In ArcGIS Pro, there are often many ways to perform a task!

2. From the options on the left, click Page Setup.

   You can always change your layout orientation to portrait or landscape. California looks fine, though, so you're going to leave it in portrait orientation. If your state looks better in landscape orientation (for example, Montana or Pennsylvania), change it.

3. Click OK to close the Layout Properties dialog box.

Resize the map frame box

4. To change the size of your map frame, click it to select it, and drag any of the grab handles (the little boxes on the corners and edges).

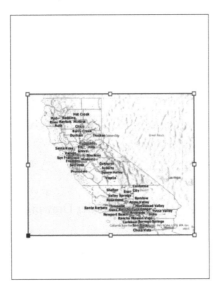

When these handles appear, you can interact with them to reposition or resize your map.

5. To change the position of your map frame, click it again to select it, and drag the box to a new position.

Insert a title

Next, let's add useful map elements to your layout. Once you add various elements, you can move them around and change their size and other properties to your heart's content. Experiment with these elements—don't worry about making them perfect at this point.

6. On the Insert tab, in the Graphics and Text group, click the plain *A* in the top row.

7. Observe that the cursor has changed. Click your layout toward the top of the page, and type **California Counties and Cities, 2022**.

   The title may not be the right size for the page. Let's enlarge it.

8. Click the text to activate the text box.

   Handles around the title text are added.

9. Right-click the activated text, and click Properties to open the Element pane.

10. To change the look of the text, click Text Symbol at the top of the pane, and expand Appearance.

    Let's start by increasing the font size.

11. For Size, choose **24 pt**. (You can also drag the handles to resize the text.)

12. Change the Font Style to **Bold**.

13. Click Apply at the bottom of the pane to apply these changes.

    If you want to try a different font or text color, knock yourself out! Just be sure to click Apply after you make any changes.

14. Click the activated text box on your layout page, and drag it so that it's horizontally centered on the page.

    > ArcGIS Pro has a handy snapping tool (blue dashed line) to help you center elements easily.

15. Click anywhere in the gray space to deactivate your title text.

Insert a north arrow

16. In the Map Surrounds group, click the North Arrow down arrow for a selection of north arrow choices.

    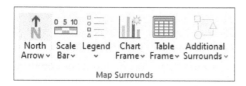

17. Choose a north arrow that you like.

18. When the north arrow choices disappear, drag a box on the layout where you want your north arrow to appear.

19. Click anywhere in the gray space to deactivate the north arrow.

Insert a scale bar

20. In the Map Surrounds group, click the Scale Bar down arrow for a selection of scale bar designs, and choose a scale bar you like.

21. Drag a box on the layout where you want your scale bar to appear.

22. Click anywhere in the gray space to deactivate the scale bar.

Insert a legend

23. In the Map Surrounds group, click the Legend down arrow for a selection of legend designs.

24. Choose Legend 1.

25. Place the legend where you want it to appear.

26. Make sure the legend is still activated. In the Element pane, under Legend > Title, uncheck the Show box.

    The north arrow is not labeled "North Arrow." The scale bar is not labeled "Scale Bar." A legend is understandable without labeling it "Legend."

27. Click anywhere in the gray space to deactivate the legend, and take in the cool new elements in your layout!

    Did you notice as you added the map elements to your layout that they were also being listed above the map frame in the Contents pane? In the Contents pane, you can turn elements on and off, click them to activate them, or right-click them for access to loads of other options, including element properties. Good to know.

Chapter 2: Creating reference maps and layouts

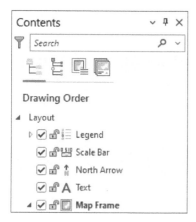

You're getting the hang of this now. Feel free to move these elements around the layout until you like the way they look. You'll work more with the layout elements in later chapters to make sure your maps are attractive and useful.

Check out how your map layout turned out.

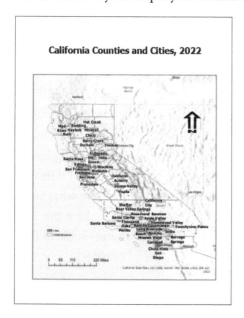

28. At the upper left of the ArcGIS Pro window, click Project, and click Save Project As.

29. Browse to your C:\GIS20\Chapter02 folder, name the project after your state, and click Save.

30. Close ArcGIS Pro.

A good map is clear to the person looking at it and communicates what the creator of the map wants to say. In this chapter, you learned how to use colors, transparency, label placement, and other tools to help you communicate. You learned how to create a layout and add a north arrow, scale bar, legend, and other useful information. In the next chapter, you'll learn how to bring data from a spreadsheet to life in a map.

Just because you can doesn't mean you should

You can add many elements to your map, but you should use them sparingly. Yes, you may have been told every map needs a title, a north arrow, a legend, an author credit, a source credit, and a print date, among other things. We hereby release you from any such requirement. Add whatever elements you need—but nothing more, because the more you add, the more distractions you add to what you want viewers to focus on. For example, if your map title is *Park Locations Downtown*, you may not need a legend with that same information. Or if your expected viewers are familiar with an area, you may not need a scale bar or north arrow.

If you're creating a map for worldwide use, where the scale isn't obvious to much of your audience, by all means include a scale bar and perhaps show both miles and kilometers if necessary. If you need a subtitle on your map to make it clear, add one. Add whatever you need—and nothing more.

As long as we're on the subject of "less is more," please refrain from seeing how many fonts and colors you can use in your map. Don't label every street if not needed. Don't add a table that isn't helpful to someone understanding what you're presenting. When we discuss basemaps later in the book, you'll see how the sparse light-gray canvas basemap is one of the most useful, whereas some of the more visually exciting basemaps are best reserved for special purposes.

Yes, we just told you to do less work. You're welcome!

USER STORY
Aligning property boundaries with GIS

In Newport Beach, California, the zoning code contains more than 300 pages of complex development standards for a multitude of subdivisions that have been developed since the city's incorporation in 1906. The city's zoning code also includes standards for building setbacks—the distance a structure or part of a structure is set back from the property line.

The city has varying setback standards for parcels based on factors such as the width and shape of a lot, how close it is to the harbor, and its adjacency to an alley.

The city uses GIS, including detailed sets of reference maps, to overcome the challenge of creating and editing the geospatial dataset for setbacks. With GIS they can accurately align setback boundaries with property boundaries.

This story originally appeared as "Setback Mapping in Newport Beach, California, Gets an Update" by Jordan Baltierra in the Winter 2020 issue of *ArcNews* and can be accessed at links.esri.com/Setbacks.

The Map Viewer app allows users to visualize and analyze complex development standards in Newport Beach subdivisions, particularly the use of setbacks, which determine the distance a structure is set back from the property line.

Credit: City of Newport Beach, Esri, VertiGIS (formerly Latitude Geographics). Data source: City of Newport Beach.

CHAPTER 3

Preparing your data

In chapter 1, you downloaded three shapefiles from the US Census Bureau (states, counties, and places). In this chapter, you'll download a data table from the census and edit it so that you can join it to one of your feature layers in chapter 4. This is a common workflow in GIS. Whether you get tabular data from the census or from another source, it's good to know the fields in your table and how to make any changes you need to match a field in an existing feature layer. The table that you'll download from the census contains age and sex demographics for each county.

Many sources of data are available, much of it free like census data. These sources may have data by census tract, zip code, city, city council district, county, or other geographic division. You'll want to convert the data into a format that GIS software can read to make it useful to you. Combining tabular data with geographic areas makes your data more useful than just storing it in a spreadsheet, as you'll see in the next few sections.

Let's get to it.

Why doesn't the US Census Bureau website look like my book?

The Census Bureau website changes from time to time—meaning often! Although the primary purpose of the census every 10 years is for congressional districting, the bureau's information is the go-to place for many of us. The site contains a vast amount of information useful to policy makers, researchers, planners, teachers, and your typical resident looking for authoritative information.

The problem is trying to write specific instructions for finding and using that data while the bureau is adding more data and working to make it more accessible. We could lament the lack of consistency, but we're thankful for the continual stream of new data and tools. If you notice discrepancies between the instructions or images in this book and what you encounter online, don't worry. You'll still be able to successfully perform the steps and learn quite a bit!

Download census files
Navigate the Census Bureau site

1. In a browser, go to **data.census.gov**.
2. Under Explore Census Data, click Explore Tables.
3. In the Filters section, click Topics, and click Populations and People.

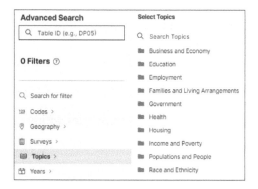

Setting a filter narrows the search.

4. Check the box for Age and Sex.

   Age and Sex is added to the list of filters on the left.

   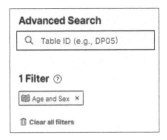

5. In the Filters section, click Geography, and click County.

   The counties may need a few moments to load.

6. Click the name of the state you used in chapter 1—we're using California.

7. When the list of counties loads, check the box for All Counties within California to add another filter.

8. In the Filters section, click Years, and click the box for 2020 to add another filter.

   All three filters are now listed under Filters (Age and Sex, All Counties within California, and 2020).

   Your search returns many choices, but the one you want is the item at the top of the Tables list.

9. Click S0101 Age and Sex.

Chapter 3: Preparing your data

Download the data table

1. At the top of the search results, click Download Table Data.

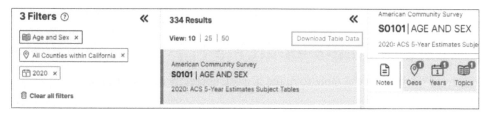

   A list of files opens that can be selected for download.

2. Check the box for S0101 Age and Sex. Above it, click Download.

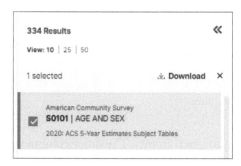

3. In the Select Table Vintages to Download dialog box, click Download.csv.

   A zip file of the comma-separated values (CSV) file downloads to your computer.

Extract and save the downloaded file

1. In File Explorer, browse to the Downloads folder (or your default download location) to locate the zip file.

   A zip file needs to be unzipped, so you'll need to extract the files.

2. Right-click the zip file, click Extract All, and save the file to C:\GIS20\Chapter03.

   The extracted files now reside in a folder in your specified location.

3. Double-click the file with "Data" in the name to open it in Microsoft Excel. (If necessary, click Enable Editing.)

You'll notice a huge number of fields in this spreadsheet. (Fields are the category headings across the top.) All those Margin of Error fields are difficult to read clearly, but the Census Bureau is thorough about ensuring data quality. We can all appreciate that.

Prepare the table for ArcGIS Pro

You need to clean up the table to make it work correctly in ArcGIS Pro.

1. In the Microsoft Excel spreadsheet, change the name of the Name column to **County**.

   Hint: Double-click a field header and type over the existing text.

2. Change the name of the S0101_C01_001E column to **Population**.

   Hint: Resize the field headers to see all the text. The S0101_C01_001E column should be directly to the right of the new County column.

   This column contains the total population for each county. The other column you want is the number of people age 65 and older.

3. On the Microsoft Excel ribbon, on the far right, click the Find & Select tool > Find. In the Find What text box, type **S0101_C02_030E**, and click Find Next. Close the Find tool.

   The correct field is highlighted in your spreadsheet. Use the tool again if your highlight disappears.

   If you expand the column width of the field you just highlighted, it will say, "Estimate!!Percent!!Total population!!SELECTED AGE CATEGORIES!!65 years and over" near the top. This column consists of the percentage of people age 65 and older in each county.

4. Change the column name from S0101_C02_030E to **Seniors**.

   You'll copy the Seniors column to the left side of the table to see it better.

5. Right-click the field column header, and click Copy.

6. Scroll to the left, right-click column C, and click Insert Copied Cells.

Chapter 3: Preparing your data

Your Seniors field is now in column C.

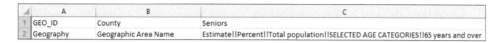

Next, you need to alter the Geo_ID field (column A) so that it matches a field in the county shapefile.

7. Click the top of the A column to select it, click the Find & Select tool, and click Replace.

The Find and Replace dialog box appears. You're going to replace some of the digits in Geo_ID so that it matches the corresponding five-digit field in the county shapefile.

8. In the Find and Replace dialog box, for Find What, type **0500000US** (five zeros in the middle).

9. For Replace With, keep the text box empty, and click Replace All.

Did you notice how the numbers in the Geo_ID column all became right aligned? That's because you removed "US" from each cell in that column. In Microsoft Excel, left aligned indicates a text field. You want that field to be formatted as text, so it will be left aligned.

10. At the top of the spreadsheet, click the Undo button.

11. In the Find and Replace dialog box, for Replace With, type a single quotation mark to tell Microsoft Excel that the numeric values should be considered text.

12. Click Replace All.

The Geo_ID values are left aligned this time.

13. Close the Find and Replace dialog box.

    You need to make two more changes to the spreadsheet before it's ready for ArcGIS Pro. You don't need row 2 anymore. In fact, if you don't delete it, ArcGIS Pro will see your Population and Seniors fields as text instead of numbers. You also don't need all those extra fields.

14. At the far left, in row 2, click the number 2 to highlight it, right-click, and click Delete.

15. Click column E to highlight it, right-click, scroll to the very last column to the right, press and hold Shift, click column AIE to highlight it, right-click anywhere in the highlighted area, and click Delete.

    Your table should resemble the following image. In Microsoft Excel, the little green triangles on the Geo_ID values indicate that those numbers are stored as text. That's exactly what you want.

	A	B	C	D
1	GEO_ID	County	Seniors	Population
2	06001	Alameda County, California	13.9	1661584
3	06003	Alpine County, California	29.1	1159
4	06005	Amador County, California	27	39023
5	06007	Butte County, California	18.2	223344

    All you need to do now is rename and save your worksheet.

16. At the bottom, click the ACSST5Y2020.S0101-Data tab, double-click it, and type **Seniors**.

17. On the ribbon, click the File tab, and click Save As.

18. In the Save As pane, browse to your Chapter03 folder, name the file **Seniors**, set the file type to Excel Workbook (.xlsx), and click Save.

    > You still have the original table you downloaded, in case you need it again.

19. Close Excel.

    So, all this data stuff seems a bit tedious and boring, right? But data is the most powerful part of GIS. You can do some amazing and powerful things using the right data. In the next chapters, you'll join this Microsoft Excel table to your file of counties and create a map with the data. You'll find Microsoft Excel handy for managing data in GIS, and you don't need to be a Microsoft Excel wizard to work with it. Read this interesting user story, and then let's jump to the next chapter.

USER STORY

Combating homelessness with housing

The *Los Angeles County Homelessness & Housing Map* is a data-driven, multilayered GIS planning tool to help policy makers respond effectively to homelessness in LA County. The tool shows both interim housing—such as shelters—and supportive housing that exists and is being developed, including sites under construction and in the planning and development phases. The tool lets users view these developments geographically and in the context of recent point-in-time count data. It illustrates gaps between where the housing needs are and where projects exist or are being developed. In addition to helping policy makers, the tool can be used to inform the public and community leaders about the details of this mobilization to combat homelessness in the county.

You can explore this interactive web app at links.esri.com/Homelessness.

This GIS web app helps city officials and planners identify neighborhoods and other areas in Los Angeles County where homelessness hot spots are located. With this information, officials can plan where additional housing and shelters are most needed.

Credit: County of Los Angeles. Data source: County of Los Angeles, California State Parks, Esri, HERE, Garmin, SafeGraph, FAO, METI/NASA, USGS, Bureau of Land Management, EPA, NPS, County of Los Angeles Department of Public Works.

CHAPTER 4

Joining tables to GIS data

With points, lines, and polygons, you can make some nice, useful maps. But you would only be scratching the surface of what GIS can do. The power of GIS is in the data. In the previous chapter, you downloaded data from the US Census Bureau and modified it to provide the information you need in a form GIS can understand. The next step is to connect that tabular data to geography.

Think how often you have tables of location-based data. It may be census tracts, states, counties, cities, businesses and stores, schools, or even street trees, such as an inventory of tree types and conditions. The process of connecting that data to geographic data is called joining.

Joining can be a bit tough because it's easy to make mistakes in setting up the data. Luckily, it's easy to undo a join and start over. We'll walk you through how it works, and you'll impress your colleagues with your new skill. As we like to say, it's so easy, even a manager can do it.

Join a stand-alone table to a shapefile

Add data to ArcGIS Pro

1. In ArcGIS Pro, start a new map project. Name it **Ch4_join**, and for Location, browse to C:\GIS20\Chapter04. Click OK to save the new project.

2. On the Map tab, in the Layer group, click the Add Data button.

3. Browse to C:\GIS20\Chapter01. Add the CaliforniaCounties.shp file. If you used a different state for your earlier project, add that file.

4. Click Add Data again, browse to the Chapter03 folder, and double-click the Seniors.xlsx file. Wait a moment for the Seniors$ worksheet to appear, select it, and click OK.

   > If you didn't rename the worksheet in chapter 3, it will still be called Sheet1$. The "$" in the name denotes a worksheet.

   In the Contents pane, the Seniors$ table is now listed at the bottom, under Stand-Alone Tables.

Find columns with matching data

1. In the Contents pane, right-click the Seniors$ table, and click Open.

   The table opens underneath the map.

   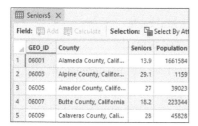

   To join data, the two sources must each have a column with data that matches a column in the other source. The columns can be called different names, but the data in the rows must match exactly. Let's look for a column with matching data in our two sources.

   In the previous chapter, you prepared the Geo_ID field in the Microsoft Excel spreadsheet. You'll look for a field in the shapefile attribute table that matches.

2. To open the attribute table for the CaliforniaCounties shapefile, in the Contents pane, right-click CaliforniaCounties, and click Attribute Table.

Two tabs now appear at the top of the table view.

3. In the CaliforniaCounties attribute table, analyze the fields to see whether one of them is similar to the Geo_ID field in the Seniors$ table.

The GeoID field in CaliforniaCounties and the Geo_ID field in Seniors$ are similar in length and data type. Both fields have data values that are left aligned, indicating that they are both text fields.

Remember: A text field can be joined only to another text field, and a numeric field can be joined only to another numeric field. If any join procedure fails for you, a mismatch of field types will probably be the reason. Check out the next section, "Numbers that Aren't Really Numbers" for a better understanding.

4. Click the *X* on each tab to close the attribute tables.

Numbers that aren't really numbers

When is a number not a number? When that number is text! Wait, what? Numbers can't be text! They most certainly can. Consider these examples:

- Telephone numbers
- Social Security numbers
- Parcel numbers
- Identification numbers
- Postal codes
- Census tracts

What these types have in common is that the digits have no mathematical meaning. You can't subtract one postal code from another. One phone number isn't larger than another. You could use letters in place of the digits, and the combination would work just as well. In fact, many countries use letters for their postal codes.

Mathematically, 02860 is the same as 2860, but a zip code of 02860 (Providence, Rhode Island) isn't the same as 2860. A parcel number of 1224-030-002 is certainly not the same as 1,224,030,002. Here's why that matters: when you try to join data in which zip codes in a GIS are text and zip codes in a spreadsheet are formatted as numbers, the join will fail. Both values need to be the same format, and if the data is really text, the format should be text.

It's easy to tell whether the format is text or numeric. In your attribute table, notice the left and right alignment of each field. In the image, the Block Group FIPS field is left aligned and is therefore text. The 2020 Total Population and Area in Square Miles fields are right aligned and are therefore numbers.

Block Group FIPS	County	State	2020 Total Population	Area in square miles
061110003021	Ventura County	CA	2631	14.74
061110003022	Ventura County	CA	2715	24.92
061110003023	Ventura County	CA	1092	0.2
061110003031	Ventura County	CA	1654	0.15
061110003032	Ventura County	CA	1726	0.41
061110003033	Ventura County	CA	1814	13.89
061110003041	Ventura County	CA	913	16.82

Now you know when numbers are—or aren't—numbers and how to tell the difference.

Join the two tables

When you perform a join, if you apply a selection to any of the records in either table, the join will be performed only on the selected records. It's good practice before performing a join to make sure that specific records haven't inadvertently been selected.

1. On the Map tab, in the Selection group, click Clear to clear any previous or inadvertent selections.

2. In the Contents pane, right-click the CaliforniaCounties layer, click Joins and Relates, and click Add Join.

    The Add Join dialog box appears. A dialog box lets you set parameters before running a tool (the Add Join tool, in this case). ArcGIS Pro often autocompletes parameters with settings, saving you time.

    For the Input Table parameter, the CaliforniaCounties layer is autocompleted because it's the layer you right-clicked. For the Join Table parameter, Seniors$ is autocompleted because it's the only stand-alone table in your project.

3. For the Input Join Field parameter, click the down arrow, and choose GEOID.

   ArcGIS Pro assumes that the Geo_ID field in the Seniors$ table is the matching field. But you should always verify that the correct field was entered before you run any tool in ArcGIS Pro.

   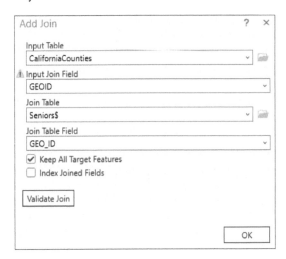

   Uh-oh. A triangular warning icon appears next to Input Join Field.

4. Click the warning to read it.

   According to the warning, ArcGIS Pro recommends indexing the Input Join Field to perform your join efficiently. Indexing can improve joins and other tasks.

5. Close the warning box.

6. In the Add Join dialog box, directly above the Validate Join button, check the Index Joined Fields check box.

   The warning icon disappears.

Chapter 4: Joining tables to GIS data   47

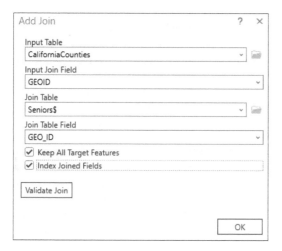

7. Click OK to run the Join tool.

Verify the join

Joins don't show a progress bar or notify you that the operation was successful. You'll need to check the attribute table to verify that the join worked.

1. Open the attribute table for CaliforniaCounties.

2. Use the slider at the bottom to scroll all the way to the right.

This table is considered the "destination" table because it's the table in which the source fields in Seniors$ were added. The Geo_ID, County, Population, and Seniors fields from the Seniors$ table have been added to the CaliforniaCounties table. Your join was successful!

3. Close the attribute table.

Create a shapefile

At this point, your join exists only within this ArcGIS Pro project. To use the joined layer in another project, for instance, you must save a new shapefile that includes the Seniors$ fields. Saving a new shapefile with those fields makes them a permanent part of the file.

1. In the Contents pane, right-click the CaliforniaCounties layer, click Data, and click Export Features.

   | Data | > | Export Features |

   The Export Features dialog box opens.

2. For Output Feature Class, browse to C:\GIS20\Chapter04, and save the file with the name **CountiesAge**.

3. Click OK.

   > *If you get an error here, you may not have deleted all the extra fields in the Seniors.xlsx file. Thought we'd never know, huh? You'll need to go back to those steps in chapter 3, repeat the process, and try this step again.*

   ArcGIS Pro adds your new CountiesAge layer to the Contents pane. If you're skeptical, you can open the attribute table and verify that all the fields are there.

4. You no longer need the original CaliforniaCounties shapefile or the Seniors$ table, so right-click each of them in the Contents pane, and click Remove.

5. Save your project and close ArcGIS Pro.

   Congratulations! You've joined a stand-alone table to a shapefile and saved it as an entirely new layer. But you may be asking, "Why did I do that?" In the next chapter, you'll use those added fields to create a map. Can't wait to get to it? Check out our user story, and then proceed to the next chapter. Otherwise, take a break—you've earned it!

USER STORY

Highway information joined for Wyoming travel map

GIS professionals use joins all the time. But they sometimes don't mention the various operations they perform behind the scenes in their GIS work. In this user story, a standard street line layer is transformed into an amazing source of information by joining other information to it.

Wyoming highways serve as major corridors for commercial truck traffic—particularly on Interstate 80, which runs through the southern part of the state. It's one of the busiest routes in the United States for moving freight coast to coast. Truckers, tourists, and locals alike can navigate Wyoming's often unpredictable highways with the help of information from the *Wyoming Travel Information Map*, built by the Wyoming Department of Transportation. The interactive map displays the current road conditions, construction areas, and advisories. On an average day, the map gets 170,000 visits—and that number can climb to four million when there's a big storm.

You can explore this interactive web app at links.esri.com/WyomingTravel.

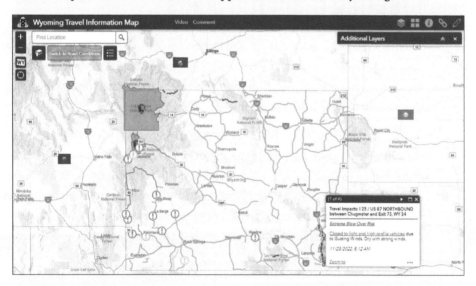

The *Wyoming Travel Information Map* tracks commercial transportation lines across the state and allows businesses and travelers to see where weather conditions, construction areas, and roadblocks may hinder shipping and travel.

Credit: State of Wyoming. Data source: State of Wyoming, Esri, USGS/NRGC, HERE, Garmin, FAO, NOAA, USGS, Bureau of Land Management, EPA, NPS, US Census Bureau.

CHAPTER 5

Creating thematic maps

Thematic maps focus on a topic and symbolize features based on quantitative data (for example, counts or percentages). Some examples of thematic maps are demographic maps, election maps, and disease transmission maps. You can undoubtedly think of many more.

In this chapter, you'll use the CountiesAge shapefile (which now contains the census data about seniors) from chapter 4 to create a thematic map by symbolizing the percentage of seniors across different counties. Why might that be important? Seniors are considered a "sensitive" population. Knowing where there are large concentrations of seniors may be important for emergency response or evacuation efforts or for examining access to health care.

Create a thematic map by symbolizing layers

1. In ArcGIS Pro, create a new map project, name it **Ch5_thematic**, and save it to C:\GIS20\Chapter05.

2. Click Add Data, and add the CountiesAge file from your Chapter04 folder.

3. In the Contents pane, right-click the CountiesAge layer, and click Symbology.

4. In the Symbology pane to the right, click the down arrow for Single Symbol.

   Initially, you'll notice many choices for symbolizing your layers. You can use the scroll bar to read the names and descriptions of even more.

   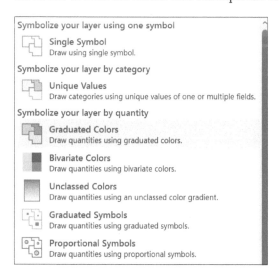

   You want to symbolize your layer by quantity. Several choices are available to do so, such as Bivariate Colors, Unclassed Colors, and Graduated Symbols, but you'll start with the basic option.

5. Click Graduated Colors.

   Graduated Colors allows you to symbolize using colors that blend continuously into one another. This color scheme lets you indicate the percentage of seniors in each county.

Refine the ways you represent your data
Change the color scheme

ArcGIS Pro assumes that you may want to symbolize the first numeric field and picks a color scheme for you, but in this case, it's not the numeric field you want to map. You want a different color scheme, too.

> *ArcGIS Pro picks the field and colors to get you started visualizing your data, but don't worry if you need to make changes or if your layer has different colors from those shown in the image.*

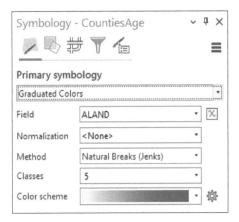

Let's start by choosing the correct field.

1. For Field, use the down arrow to choose Seniors.
2. For Color Scheme, choose the Blue-Green (Continuous) color scheme.

> **Hint:** Check the Show Names box or point to a color scheme to see the names.

For now, you'll keep the default values for the other options. Your Symbology pane should resemble the image.

Chapter 5: Creating thematic maps

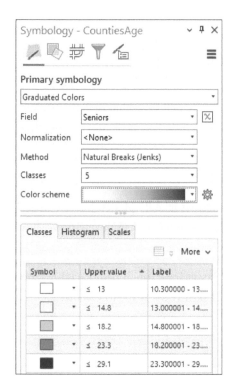

Change the look of the percentages

We know that the Seniors field in the table contains percentages of seniors in each county, but the percentages look confusing in the Contents pane because of all those extra decimal places. That's all right—we can fix that!

| *Your numbers will be different if you used any state other than California. Not a problem.*

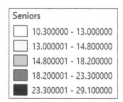

1. At the top of the Symbology pane, click the Advanced Symbology Options icon.

    | *Remember to point to each icon to see a ScreenTip that tells you the name of the tool.*

2. Click Format Labels to expand it.

3. For Category, click the down arrow to choose Percentage. Observe how the values change automatically in the Contents pane.

Your numbers already represent percentages, so you'll keep that option checked under Percentage. But you still have an unwieldy number of decimal places that need to be fixed.

4. Under Rounding, change Decimal Places to **0**, and press Enter.

You'll almost never want six decimal places. One is usually plenty, but use the number of places that's appropriate. Don't just accept the default setting without thinking about what your map needs.

In the Contents pane, the layer and its five classes look much better, right? Good job. Rounding the percentages is an essential skill you'll want to use often.

Change the class breaks

1. At the top of the Symbology pane, click the Primary Symbology icon, and observe that this symbolization uses five classes.

2. For Classes, click the down arrow, and change the value to **4**.

3. In the Symbology pane, click More.

4. Click Show Statistics to see statistics for your data.

Statistics give you a quick idea of how your data is distributed and can help you make good decisions about how to represent the data.

Under Primary Symbology, Method is set to Natural Breaks (Jenks). You'll learn more about classification methods later, but you'll keep Natural Breaks for now.

The classes and method we're using for now aren't necessarily the best. But you're learning how easy it can be to change the representation of your data.

Change the listing order of the values

The legend shows the highest values at the bottom. Suppose you wanted to reverse the listing order. Let's reverse those values so that the highest percentage of seniors is the darkest value and appears at the top.

1. In the Symbology pane, click More, and click Reverse Symbol Order.

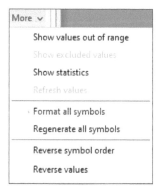

Reversing the symbol order puts the darkest value at the top, but the highest numbers are still at the bottom.

2. Click More again, and click Reverse Values.

The highest values are now at the top of the legend and have the darkest color. This color scheme makes clear that the high percentage of seniors is the theme of the map.

Change the basemap

By default, the basemaps listed at the bottom of the Contents pane are World Topographic Basemap and World Hillshade. The labels on World Topographic Basemap look odd. Let's change the basemap to something more suitable.

1. On the Map tab, in the Layer group, click Basemap.

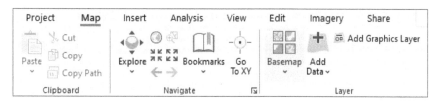

2. Observe the variety of basemap choices. For this map, click Light Gray Canvas.

   The map also has some distracting labels, but we can turn them off.

3. In the Contents pane, uncheck the Light Gray Reference layer box to turn off the basemap labels.

   The map now has visual context with the other states, and labels no longer create a distraction.

Esri basemaps

It's helpful that ArcGIS Pro starts with a default basemap (World Topographic Basemap or World Hillshade). Basemaps are the contextual background on which you can add all your own GIS layers. It wasn't always like this, however. In the past, GIS professionals had to build their own local basemap with administrative boundaries, street and place-names, bodies of water, and so on, for their own area. Today, Esri creates basemap layers for users worldwide. They aren't saved locally but stream to the software from Esri servers. (If your internet connection isn't great and your map draws slowly, you can turn off the basemap layer temporarily to improve performance.)

Esri offers many basemap types, ranging from the practical to the whimsical. Check out the different basemaps so that you're aware of this range. Always ask yourself, "Which basemap best enhances the message of my map?" Sometimes the answer is that none of them do, in which case you can simply turn off the basemap layer. Easy peasy!

Add labels

ArcGIS Pro assumes what you want to label. Sometimes, those choices are suitable for your project, but not always. Let's check out the labeling properties to be sure.

1. In the Contents pane, right-click the CountiesAge layer, and click Label.

2. Make sure the CountiesAge layer is selected in the Contents pane, and click the Labeling tab on the ribbon.

Chapter 5: Creating thematic maps

In the Label Class group, the Name field is being used for labeling. Looks good!

3. In the Text Symbol group, click the down arrow to expand the gallery of preset styles you can use for improving your labels.

4. In the gallery, under Scheme 2, click Populated Place.

5. In the Text Symbol group, change the size to **8 pt**.

   That looks better! The labels are more readable. The next step is to showcase your map on a page layout.

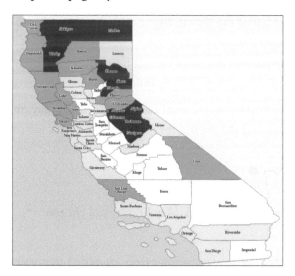

Create a layout

You created a layout in chapter 2, so let's see what you remember.

1. On the ribbon, click the Insert tab. In the Project group, click New Layout.

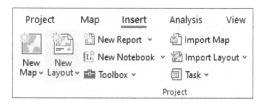

2. For this map, choose the Portrait Letter size (8.5-by-11-inch).

   The view shows a blank page on the Layout tab next to the Map tab.

The ups and downs of guides

Some ArcGIS users rely on guides to align the map and other elements and help structure their page. But cards on the table here: we personally don't like guides. Why? They're great for snapping elements into position, but they don't discriminate—they snap anything and everything, including elements that you don't want snapped. They just can't help themselves! But by all means, try guides for yourself and make your own call. Here's how you would add guides to your current layout:

1. Find the rulers on the top and left of the layout. Right-click the 0.5-inch tick mark, and click Add Guide.
2. Add three more guides so that your layout has a total of four guides. Position the guides 0.5 inches in from the top, bottom, left, and right.

If you find that guides help you control the movement of elements in your layout, feel free to use them. But recognize that they're not for everyone.

The first thing you need to add to the page is a map frame.

3. On the Insert tab, in the Map Frames group, click Map Frame, and choose your map.

The map isn't added automatically. You may remember from chapter 2 that you first need to tell ArcGIS Pro where on the page you want the map to go.

4. Click and drag to draw a box to define the placement. (You can resize or move it later to make room for your title.)

The entire state may not be visible anymore. We can fix that!

5. In the layout, click once inside the map frame to make it the active (selected) element.

   You should be able to see little boxes, or grab handles, on the sides and corners.

6. Right-click inside the map frame, and click Activate.

The layout changes slightly so that everything but the map turns gray.

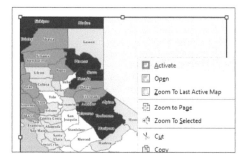

7. In the Contents pane, right-click the layer, and click Zoom to Layer.

   Zoom to Layer centers the map in the frame. Once the map is activated, you have access to all the Explore tools (such as Zoom and Pan) that you used in chapter 1.

8. Pan and zoom to adjust your layout. When you've finished, at the upper left of the layout frame, click Layout to deactivate the map.

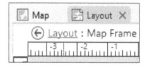

Insert a title, north arrow, and legend

Does adding a title, north arrow, and legend sound familiar? You added these same elements to your first map in chapter 2. Let's see what you remember as you practice these skills again on this thematic map.

1. Add a title, north arrow, and legend to your layout.

   > **Hint:** If you need a reminder, you can peek back at the chapter 2 exercise, "Modify and Insert Layout Elements," to review these skills.

2. After you've added a legend, right-click it, and click Properties.

3. In the Element pane, under Options, expand Legend Items, and click Show Properties.

4. Expand Show, and uncheck the Layer Name box.

   The layout looks cleaner without the unnecessary layer name.

   Your final map should resemble the following image:

   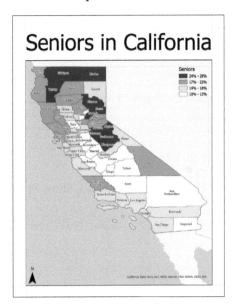

   Such a fantastic map deserves to be saved!

5. Using the skills you've learned, save and close your project.

Now that you can make a thematic map, you can pack a tremendous amount of information into a map that viewers can understand. Using color symbology, you can effectively say, "Here's where we need to concentrate our efforts," or "Here's where something worthy of further study is happening." That map may take only seconds to provide that information or may be a source of knowledge that viewers can take in over time and refer to repeatedly.

Check out the following information about "lying" with maps—something you'd never want to do, right?

Lying with maps?

Mark Monmonier wrote the popular book *How to Lie with Maps* in 1991. It's a fascinating book for anyone who uses maps, but it's perhaps most valuable for people who create maps. The following examples represent some of the myriad ways that maps can be structured to "lie"—or tailor a message for better or worse:

- Using a classification system for a color scheme that intentionally minimizes or maximizes differences.
- Using a map projection that intentionally distorts areas to make them look less or more significant.
- Using raw numbers instead of percentages—or the other way around. Showing crime as a number of incidents per thousand residents is different from showing the total number of crimes. Both numbers are useful, and both could be used deceptively.
- Comparing maps where one is a raw number and the other is a rate. We were the recipients of such a map once where it was being used to show the need for tens of millions of dollars for an area. Misleading? You bet.
- Using red or yellow to make an area look scary. Those colors indicate danger in many parts of the world. Do you really intend to make an area look scary? Red stands out and looks like a larger area than the same area in a color such as dark green.
- Using area to show people. Large areas often contain far fewer people than smaller areas. Most people in the world live in cities, so showing a map comparing areas can be misleading.

These are just some examples. Read the book for more. But wait until you complete this book, of course. Focus!

USER STORY

Cartographer visualizes how to lie with thematic maps

In *Thematic Mapping: 101 Inspiring Ways to Visualise Empirical Data* (Esri Press, 2021), the cartographer Kenneth Field writes, "Maps are designed to make lies appear truthful, misinformation respectable, and to give an appearance of fact to pure illusion." Field illustrates his point by showing a map that the then president Donald J. Trump used to celebrate his first 100 days in office after his election victory in November 2016. Trump pointed to a US map to show how he turned the country red because the blue on it was minimal. People criticized this map as obfuscating the true picture, but, in fact, the data was correct. "But it was a map that painted a specific picture of the results through the choices the mapmaker had made. It was a map that had a lot of Republican red on it and which perfectly illustrated the apparent strength and extent of his win using a certain thematic mapping technique (a choropleth map) coupled with a version of the data that emphasised his support (percentage share of the vote by county). Any Republican victor would show the same type of map with the same design choices."

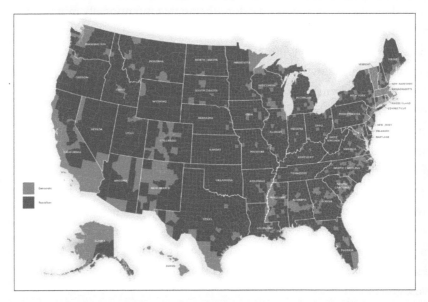

Maps such as this one of voting patterns can accurately reflect the underlying data but still provide a misleading picture of an important topic. Symbology and other factors in a map can be tailored to emphasize or de-emphasize certain aspects, depending on the map creator's needs.

Data source: US Census Bureau. © 2016 M. E. J. Newman.

CHAPTER 6

Geocoding

One of the most powerful functions of GIS is the ability to match addresses to their locations on the earth. This process is called geocoding. In seconds, ArcGIS Pro can match thousands of addresses and create a point for each one on a map. Before long, you'll quickly take geocoding for granted (which is fine), but for now just imagine how long it would take you to match even 20 addresses in a table to their correct positions on a map. Neither fun nor productive.

Geocoding can be useful for many types of data, such as crimes, schools, agencies, business licenses—anything with a location that you want to show on a map.

Geocoding can be used to locate one address on an online mapping site (yes, that's GIS working in the background of your mapping app). It can save users time by locating multiple addresses in a table. It can also locate thousands of addresses at one time, something users would probably never attempt to do manually.

How does GIS accomplish this magic? By using something called an address locator.

What's an address locator?

An address locator is a file that tells the geocoding function how to match tabular address data to a feature, such as streets. It specifies which fields in a table have certain types of information in them. For instance, is the street address information part of a field called "Address" or "Location" or maybe "CrimeScene"? The address locator tells the geocoder which field to use for the street address. It sounds simple but can be complex: data can be geocoded in many ways.

Originally, individual users had to use their own street GIS data and create their own address locator. Although you can still make your own address locator, Esri provides a versatile address locator for you. It's called the ArcGIS World Geocoding Service. It consumes credits from your ArcGIS Online account, but they're minimal, unless you're working with a huge number of addresses. Even then, it's a bargain. We'll make sure that it's a minimal amount for this lesson. You'll learn more about credits in chapter 9.

Use an address locator to geocode data

Add tabular data to a project

1. In ArcGIS Pro, start a new map project. Name it **Ch06_geocoding**, and save it to your Chapter06 folder.

2. Click the Add Data button. From the Chapter06 folder, add Facilities.csv.

    Facilities is added to the Contents pane under Stand-alone Tables.

    The table lists service facilities in the San Jose, California, area with all the address fields you need (Address, City, State, Zipcode). Let's open the table to check out those fields.

3. Right-click Facilities, and click Open.

	NAME	ADDRESS	CITY	STATE	ZIPCODE	FACILITY_TYPE
1	Saratoga Retirement C...	14500 Fruitvale Ave	Saratoga	CA	95070	Retirement Community
2	Absolute Senior Soluti...	4125 Blackford Ave Sui...	San Jose	CA	95117	Senior Services
3	Home Instead	1006 Stewart Dr	Sunnyvale	CA	94085	Senior Services
4	Alma Senior Center	136 W Alma Ave	San Jose	CA	95110	Senior Center
5	Seniors N More In Ho...	1900 Camden Ave	San Jose	CA	95124	Senior Services

4. When you've finished looking at the fields and data values, close the table.

Geocode the tabular data

1. In the Contents pane, right-click the Facilities table, and click Geocode Table to open the Geocode Table pane to the right.

   ArcGIS Pro provides a user-friendly workflow to walk you through the process step by step.

   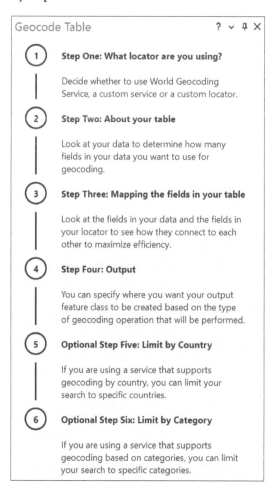

   Looks cool! Let's get to it.

2. At the bottom of the Geocode Table pane, click Start.

   Once you get the hang of geocoding, you can click Go to Tool instead to go straight to the geocoding tool.

3. In Step One ("What locator are you using?"), for Input Locator, click the down arrow, and choose ArcGIS World Geocoding Service. Click Next.

In Step Two ("About your table"), the input table was autopopulated. Before you started the workflow, you examined the table and learned that the address data was stored in multiple fields, so you'll keep the More Than One Field value.

4. Click Next.

In Step Three ("Mapping the fields in your table"), you'll specify which fields in your table to use in the fields that the address locator needs. ArcGIS Pro provided the likely fields (for example, the State address field was automatically matched to the State data field). The fields have fairly standard names, so the matching is correct. The Address, City, State, and Zipcode fields were identified and will be used to develop the locator. Does the mapping look correct? Yes, it does.

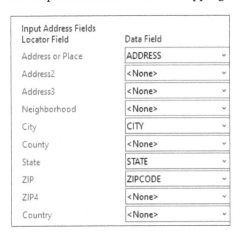

5. Click Next.

   In Step Four ("Output"), you can specify the name and location of the output layer.

6. Keep the Facilities_Geocoded name, and save the file to your default location.

7. Make sure that the box for Add Output to Map after Completion is checked.

   Checking this box automatically adds the point layer to the map when the geocoding is finished.

8. Examine the choices for Preferred Location Type and Output Fields, but keep the default values: Address Location and All, respectively. Keep all other defaults, too.

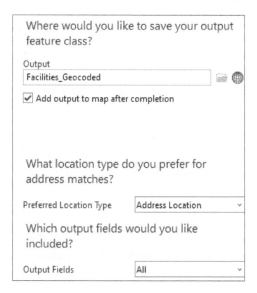

9. Click Next.

   All your data is from the United States and consists of ordinary street addresses, so you'll set those limits.

10. In optional Step Five ("Limit by country"), at the top of the list, check the box for United States. Click Next.

11. In optional Step Six ("Limit by category"), check the box for Address. Even though you aren't using them now, review the other choices so you're familiar enough with them to use them at another time.

12. At the bottom of the pane, click Finish to run the geocoding.

    Oops! The geocoding didn't immediately launch. Instead, the Geocode Table pane generated a couple of messages. The first message is a reminder that the ArcGIS World Geocoding Service consumes ArcGIS Online credits. We talked about that already, but the reminder is nice. The second message reminds you to review your workflow input. Everything looks correct—good job!

    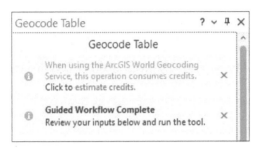

13. For the message about credits, click Estimate Credits to see how many credits this process will use.

    You'll use fewer than one credit for 20 addresses. That's a bargain!

14. At the bottom, click Run to run the geocoding tool.

Examine the geocoding results

The geocoding finishes within seconds. Three things happen. First, a status message appears with details about how many of the addresses were matched successfully and how many records per hour were geocoded. The status message provides another opportunity to match any records that weren't successfully geocoded. All 20 of your addresses (100 percent) were matched successfully, so you don't need to start a rematch.

1. Click No.

    Second, a success message appears at the bottom of the Geocode Table pane.

    Third, the new Facilities_Geocoded point layer is automatically added to the map and the Contents pane.

    Hint: If the table is still open, close it so you can view the map.

The map, with points indicating facilities, should resemble the following image.

2. Save and close your project.

   Geocoding is pretty cool, right? GIS can geocode for us so we don't have to manually locate addresses on a map. You'll probably use this skill often, so pat yourself on the back for learning such an important shortcut.

Geocoding by referencing a point layer of addresses

In this chapter, you geocoded by referencing a street line layer. That's by far the most common geocoding method, but you can also geocode by referencing a point layer. If you have a point layer of addresses, you can use an address locator for Single Field and geocode to the field in your point layer that contains the addresses. Why would you want to do that? When you geocode using a street line layer, the geocoder estimates where along a given block a particular address should be, but if you reference a point layer that already has verified precise locations for all points (each point has an address and is in the correct location), the results may be more accurate than estimating the location. Good to know!

USER STORY

Analyzing street crime using Los Angeles GeoHub

A group of students from the University of Southern California (USC) used open data from the City of Los Angeles GeoHub to build an app that better explains crime in the city over the past 10 years. The project, called A Spatial Analysis of Street-Level Crime Trends in Los Angeles, was initiated in fall 2016—less than a year after Los Angeles launched its GeoHub (geohub.lacity.org), a public platform that allows anyone to explore, visualize, and download location-based open data.

The students used GeoHub data and geocoding to assign a street to each crime. They created a mapping app that brought the crime data together with social and built-environment characteristics, incorporating variables such as unemployment rates, the presence or absence of streetlights, and proximity to metro rail and bus stops. The team then analyzed the data according to specific street segments.

This story originally ran in *ArcNews* in 2017. You can further explore the story at links.esri.com/USCGeoHub.

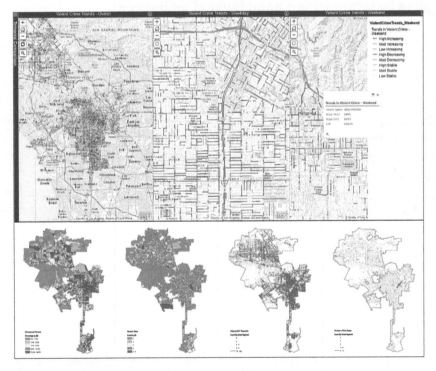

A Spatial Analysis of Street-Level Crime Trends in Los Angeles map app.

Data source: City of Los Angeles, County of Los Angeles, Bureau of Land Management, Esri, Bonnie Shrewsbury.

CHAPTER 7

Creating categorical maps

Earlier in the book, you made a thematic map showing the percentage of seniors across different counties. Thematic maps, again, are based on numeric, or quantitative, data.

Now you'll learn about the other main type of map. Categorical maps symbolize features based on categories, or qualitative data. Categorical maps can be used to symbolize polygon features by category, such as soil type, land cover, or zoning. They can be used to symbolize line features, such as street type, or to symbolize point symbols, such as different types of crime. Can you think of others?

In this chapter, you'll make a categorical map of different types of service facilities. You'll examine how symbolizing features according to their types can be valuable. It all goes back to conveying an intended message with your maps. You're probably getting the hang of that now!

Examine the layer attributes

1. In your Chapter06 folder, browse to your Ch6_geocoding folder, and open your project.

2. In the Contents pane, open the attribute table for the Facilities_Geocoded layer.

3. In the table, scroll all the way to the right to find the Facility_Type column.

   This column describes the type of service provided by each facility.

4. Close the attribute table.

5. In the Contents pane, right-click the Facilities_Geocoded layer, and click Symbology.

6. In the Symbology pane, click the Single Symbol down arrow, and choose Unique Values.

   > Symbolize your layer by category
   > **Unique Values**
   > Draw categories using unique values of one or multiple fields.

   Geocoding adds a lot of extra fields to the original Facilities table. The geocoding fields are found at the beginning of the list, followed by the original fields at the end.

7. Click the down arrow for Field 1, and choose Facility_Type from the bottom of the list.

   The Facility_Type field populates the legend in the Contents pane and the lower half of the Symbology pane. There are 10 symbol classes in that field, representing types of facilities.

   The color scheme that ArcGIS Pro applied contains colors that are too similar and points that are too small.

8. Click the down arrow for Color Scheme, and choose the top option: Basic Random.

   > **Hint:** You can see the name by checking the Show Names box or by pointing to the color scheme.

   The Basic Random colors are distinctive. Now each type of social service facility has its own color.

9. On the Classes tab of the Symbology pane, click More, and choose Format All Symbols.

10. In the Format Multiple Point Symbols pane, click the Properties tab.

11. Under Appearance, change the size to **10 pt.**, and click Apply at the bottom.

   Your results look a lot better!

12. Click the back arrow to return to the Primary Symbology options.

The All Other Values category is turned on by default. Sometimes it's good to have All Other Values turned on so you can see whether any features are missing a value in the field you're symbolizing. Your data doesn't have any other values, so you'll turn it off.

13. Click More again, and click Show All Other Values to turn off that option.

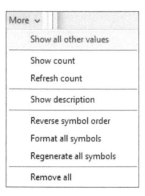

All Other Values disappears in the Symbology pane and the Contents pane. Your map should resemble the image.

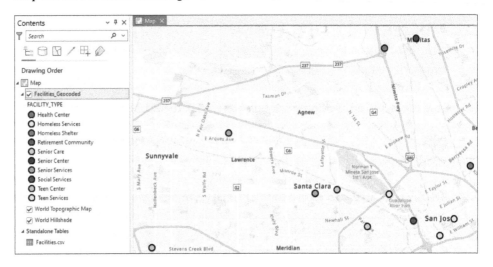

Display only selected types of facilities

You're not interested in all 10 symbol classes, so you'll eliminate some of them. As with many tasks in ArcGIS Pro, you can often perform the same task in several ways, depending on your preference. You can eliminate categories one at a time, for example, by right-clicking the name and choosing Remove.

1. Right-click Homeless Services, and click Remove.

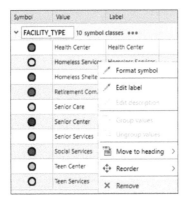

You can also remove them all and then add a few of them back.

Chapter 7: Creating categorical maps

2. Click More, and click Remove All.

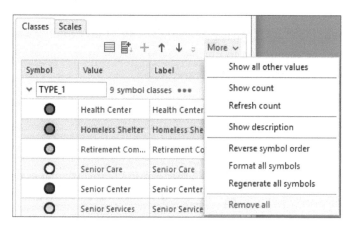

In this case, you want to view only those facility types that are related to services for seniors.

3. Still on the Classes tab, click the green plus sign. While holding down the Ctrl key, click the Retirement Community, Senior Care, Senior Center, and Senior Services items. Click OK to add them back.

> *Click the words for each category and not the symbol itself or you'll open the symbol properties. If that happens, click the arrow under the word Symbology.*

If you forget a category, you can repeat this process to get your list exactly right.

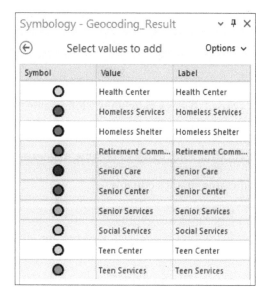

The way you approach changing your category list depends on whether you have fewer to remove or fewer to add. Either approach works. Use the approach that means less work for you! Don't forget that you want to avoid clutter or distractions on your map. You have our permission to include fewer categories as appropriate. Next, we'll show you another way to make a clearer map.

Group multiple categories into a combined category

Sometimes combining individual categories into a larger grouped category makes the message in your map clearer. You're going to combine the four categories into one combined category called Senior Services.

1. In the Symbology pane, hold the Shift key and click all four categories. Right-click and choose Group Values.

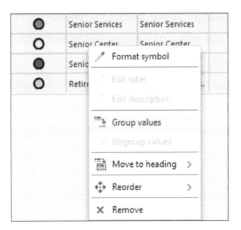

The Contents pane now has one point symbol. It has a ridiculously long name, but you can fix that.

2. In the Contents pane, carefully double-click the long name to rename it.

The text is highlighted in dark blue, indicating that it's editable.

3. Type **Senior Services.**

   > As you learned earlier, the symbology for your point may be a different color. That's OK.

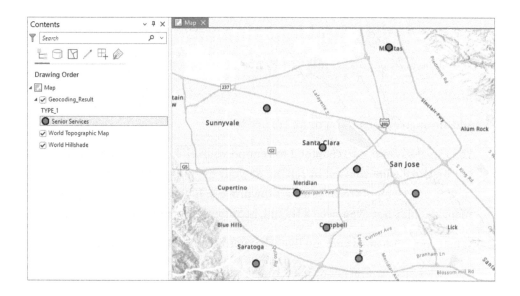

A map of qualities

You may think most maps deal with quantities—for example, demographic data such as incomes or environmental data such as air quality. But much of what we see in the world is categorical or qualitative data. Suppose you want to look at schools. You'd want to see elementary schools symbolized differently from high schools or colleges. Differences could be shown using symbols, colors, fill patterns, or label fonts for each category. A mix of these visual clues can be found in any commercial map. Street maps, too—you'll probably see a type of line for highways and another type for neighborhood streets based on size, color, and other characteristics.

All these options are available, and most of them are straightforward to use. When you're making a map, you can get fancy if you want, but you can do quite a bit without much effort through your choice of symbology and create a map that communicates what you want to say. Good maps—maps that convey precisely what you intend—don't need to take much time or cost much money.

Create a layout and add elements

You should be getting good at this part by now! Let's practice a little more.

1. Create a layout.

2. Add any map elements you think are needed.

   You can always refer to chapter 2 to remind yourself how to add layout elements, but first challenge yourself by relying on your memory. You probably remember how to add elements once you start the process. If you create a great title for your map, you may not even need a legend, because there's only one layer of data.

3. When you've finished and you like what you've done, save and close your project.

   You've learned several good skills. You've learned that you have many options for symbolizing your data to best convey your map's message. You're encouraged to explore more options that ArcGIS Pro provides. Look at your map and ask yourself, "How do I want to convey the meaning of my map? If I choose a particular element or style, will my map's theme be distorted or my message lost?" Dive into the Symbology pane and figure out how to show the features the way you want. Know the best practices but be creative! You'll definitely get better with practice.

USER STORY
Helping law enforcement track crime incidents

Crime data is classic categorical data. You've probably seen maps in which crime incidents are symbolized by the type of crime. Maps can even be symbolized by whether the crimes were misdemeanors or felonies.

In 2020, when the COVID-19 pandemic started, the Houston Police Department noticed an increase in street robberies and burglaries. To get information in officers' hands more efficiently, the department's GIS unit created a mobile app to provide current information and promote situational awareness.

The crime analyst team and the department GIS team created a dashboard that includes a map of current and historical crime incidents, allowing users to filter incidents by type and read reports in the app instead of filing time-consuming requests with a crime analyst. Information that once took analysts two hours a day to produce is now available immediately to officers in their patrol cars. Understanding where incidents occur helps the department address crime surges and allocate resources efficiently.

Visit links.esri.com/HPDCrimeApp to read more about the Houston Police Department Jugging and Sliding Dashboard.

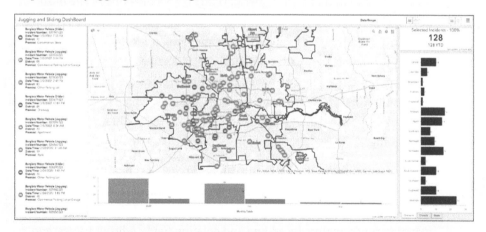

Houston Police Department Jugging and Sliding Dashboard, which shares data with officers as quickly as possible.

Credit: City of Houston, Houston Police Department. Data source: City of Houston, Houston Police Department, Esri, NASA, NGA, USGS, Texas Parks & Wildlife, HERE, Garmin, SafeGraph, MET.

CHAPTER 8

Working with data tables

Data tables? Ugh, that sounds awful, right? It's actually one of the most powerful parts of GIS. As a mapmaker, you have power because of data. If your data is limited, it's likely that your mapping options will be limited, too. But if your data is rich with relevant fields and values, your mapping options can seem limitless. The attribute data behind each of the features in your map determines how you can symbolize those features, what charts or graphs you can show, and what analyses you can perform. You're still not convinced? Keep going.

Although modifying attribute data can be done in a spreadsheet program, ArcGIS Pro has powerful capabilities for modifying data, too. In this chapter, you'll learn how to manipulate data tables in ArcGIS Pro to optimize your mapping functionality. By adding and deleting fields, editing values, and calculating values, you can strengthen your analyses and improve the effectiveness of your maps to communicate your message.

Why might all this be important? Sometimes you get data from outside sources. Unfortunately, this data tends to vary widely in quality, especially if values are manually entered. If you have the skills to clean poor-quality data, your maps will be better and their message clearer.

For now, you'll start by making changes to the data's attribute table.

Add a shapefile and open the attribute table

1. In ArcGIS Pro, start a new map project. Name it **Ch8_tables**, and save it to your Chapter08 folder.
2. On the Map tab, in the Layer group, click Add Data.
3. In the Add Data dialog box, browse to your Chapter04 folder, and click the CountiesAge shapefile you created. Click OK to add it to the map.
4. In the Contents pane, right-click CountiesAge, and click Attribute Table.

Edit data in the attribute table

In ArcGIS Pro, all data is editable at all times, so you don't need to do anything special to begin editing. This functionality is handy, but it can also be risky. We recommend backing up your data before editing. Backing up is a best practice and can save you hours of tedious repair work (we speak from experience!).

Undo

The Undo button on the ribbon (the left-curving arrow) may prove to be your best friend (we still love Add Data, too, of course). You can click the button itself or click its down arrow, which allows you to choose a specific operation to undo. Undo is the equivalent of Ctrl + Z in a Windows environment. Good to know for any moments of panic!

If you decide to undo your undo, click the Redo button to the right of Undo (the right-curving arrow). It's a handy way to get a quick before-and-after view of an edit.

Let's get to editing the attributes.

1. On the ribbon, click the Edit tab.

   Notice the Save button in the Manage Edits group. You haven't edited anything yet, so that button should be unavailable.

2. In the attribute table, scroll to find the CLASSFP field.

   Suppose you learned that the first five values should be "C1" instead of "H1."

3. Double-click the first row's value of H1 in the table.

   The cell is outlined in green, and the green edit (pencil) icon appears at the beginning of the row.

4. In the cell, type **C1**, and press Enter.

   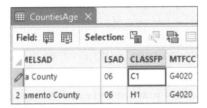

   The Save button on the Edit tab is now a color icon. ArcGIS Pro knows that you've edited something that hasn't been saved yet, so this is how it prompts you to save your edits.

5. Double-click the CLASSFP values for rows 2 through 5, and change those values to **C1**.

6. Click Save to save your data edits.

7. When prompted to save all edits, click Yes.

8. Click the *X* by the table name to close the attribute table.

   That was easy, right?

Edit data outside the attribute table on a polygon-by-polygon basis

You can also edit features directly in the map rather than in the attribute table.

1. To continue editing, on the ribbon, click the Edit tab.

2. In the Selection group, click Attributes to open the Attributes pane on the right.

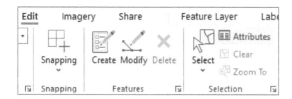

You have two tab options in the Attributes pane. Layers lets you choose one of the layers in your map. You have only one layer, so you don't need to set it now, but it's good for you to know where it is. The other tab option is Selection. The default is to select a polygon with a single click. But if you click the Select One or More Features down arrow, you can select features using several tools: rectangle, polygon, lasso, circle, line, and trace. The last three tools—box, sphere, and cylinder—are unavailable because you're not using 3D data. (Don't worry, because you'll get to work with 3D data later in the book.)

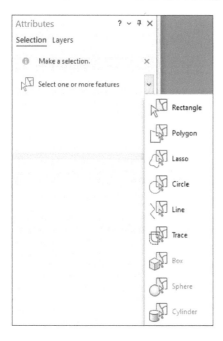

3. Click Select One or More Features.

   Your pointer changes into a selection pointer, which you can see when you hover it over the map.

4. On the map, click any of the county polygons.

   The county you selected is highlighted on the map, and its attribute information is listed in the Attributes pane.

5. In the Attributes pane, under the NAMELSAD field, click the county name.

   The county name in the table is highlighted in cyan, indicating that you can edit it.

6. Type **Co** instead of County.

7. Click inside the gray space around the word "Geometry" to deselect the text.

   The Save button is still unavailable, so your edit hasn't yet been made. Also, the Apply button at the bottom of the Attributes pane is now black instead of gray. ArcGIS Pro gives you an "out" here—a Cancel button.

8. You want to keep your edit and not cancel it, so click Apply to save it.

9. Close the Attributes pane.

   The Save button on the Edit tab is now a different color because ArcGIS Pro knows that you completed an edit.

10. Click Save, and click Yes to permanently save your edits.

    ArcGIS Pro lets you know that you have unsaved changes by displaying the Save button in color. If the button is unavailable, you can trust that you've already saved your edits.

Add a field to the attribute table

You may want to symbolize your data or perform analyses with a value that isn't in your table. You can add a new field and calculate values for it.

1. If necessary, open the CountiesAge attribute table. (If you don't remember how, refer to the beginning of this chapter.)

2. At the top of the table, click Add.

Chapter 8: Working with data tables

The Fields View tab opens to the right of the Attribute Table tab. A new field, named Field, has been added to the bottom row.

3. Double-click the word Field, and type **Percent**.

4. For the Data Type column, double-click the cell in the new row, and choose Double from the drop-down options.

The Double numeric data type offers the most flexibility. If you're not sure what to choose, Double is usually a safe bet. You can leave the rest of the columns as is. ArcGIS Pro knows that Double is numeric, so it will fill that in for you in the Number Format column.

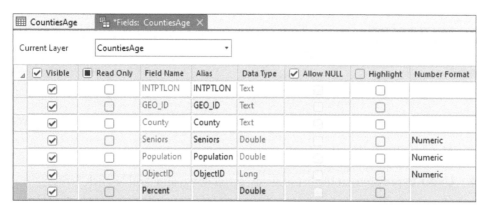

5. At the lower left of the Fields View tab, check the easy-to-miss Click Here to Add a New Field box.

6. In the new row you just added at the bottom, double-click the word Field, and type **Count**.

7. In the Data Type column, double-click the cell in the new row, and choose Double from the drop-down options. Keep the other columns as is.

8. On the ribbon, on the Fields tab, in the Changes group, click Save to save your new fields.

9. Close the Fields View tab.

Calculate values

Now that you have two new empty fields, you'll populate them with a multiplier and the total number of seniors. The Seniors field already has the percentage of seniors, but it's formatted differently. You want your Percent field in a format that, when multiplied by the total population, returns the number of seniors in each county into your Count field.

If any records are selected, Calculate Field will operate only on those selected records. You want to calculate all the records, so first check the bottom of the table to make sure that zero (0) records are selected.

1. At the top of the attribute table, click Clear to clear any existing selections.

    There are different ways to open the Calculate tool—you'll try two.

2. At the top of the attribute table, click Calculate Field.

    Opening the tool this way means you need to specify the field you want to calculate.

3. In the Calculate Field tool, for Field Name, click the down arrow, and click Percent.

    Now you'll calculate the expression.

4. Under Expression, in the list of fields, scroll down and double-click Seniors.

    The Seniors field is added to the expression box under Percent =. It's already formatted correctly, with exclamation points before and after.

5. Under the Helpers list, click the forward slash (/) division symbol to add that to your expression.

6. Type **100** in the field.

Chapter 8: Working with data tables

7. Click OK and monitor the Calculating Values progress bar while the calculations are being made.

8. Verify that the values in the Percent field in the table look correct.

   You're ready to calculate the number of seniors in each county. Let's access the Calculate Fields tool another way this time.

9. In the attribute table, right-click the Count field name, and click Calculate Field.

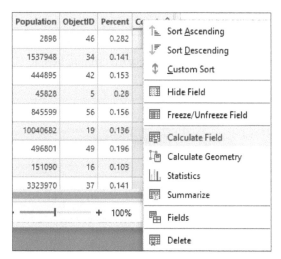

Because you launched the tool from a specific field, Count is already populated for Field Name.

10. If the previous expression is still there, clear it.

11. Under Expression, in the list of fields, scroll down and double-click Population to add it to the expression.

12. Click the multiplier symbol (*) to add it to the expression.

13. Double-click the Percent field to add it to the expression.

14. Click OK to calculate the field.

You just used the total population and the percentage of seniors to calculate the actual number of seniors in each county. Nice work!

Change the number formatting

ArcGIS Pro made the calculations, but the number formatting includes decimal places. The count is the number of people, so it would make more sense if the count were rounded to the nearest whole number.

1. Right-click the Count field, and click Fields to open the Fields view.

2. Scroll down to the bottom of the list.

3. In the Count field name row, double-click Numeric in the Number Format column.

4. Click the More button (three dots) to open the Number Format dialog box.

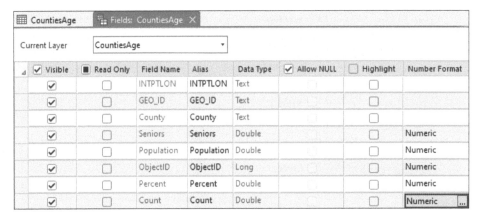

5. In the Number Format dialog box, under Rounding, change Decimal Places to **0** instead of 6.

6. Check the Show Thousands Separators box, and click OK.

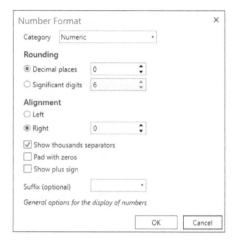

7. Click Save on the Fields tab to save the number formatting changes, and close Fields.

   The values in the Count field are better formatted for readability.

Delete fields

The CountiesAge file has many unnecessary fields in the table. You can get rid of many of these fields because you won't be using them. There are two ways to remove the unneeded fields: (1) you can permanently delete them from the attribute table, or (2) you can hide them from displaying in the attribute table view. The hiding option is the safer choice because it doesn't permanently remove any of the data fields.

First try permanently deleting a field.

1. In the attribute table, right-click the CSAFP field, and click Delete.

2. Click Yes to confirm that you want to permanently delete the field.

   Now try hiding the remaining fields to prevent them from displaying in the attribute table.

3. In the attribute table, click the Options menu (three stacked lines) in the upper right, and click Fields View.

The Fields view opens.

4. Uncheck the Visible box for three of these unneeded fields: FUNCSTAT, ALAND, and AWATER.

5. On the Fields tab at the top, click Save to save these changes.

6. Close the Fields view, and examine the attribute table (CountiesAge) to confirm that those fields are no longer visible.

   Hiding the unwanted fields keeps you from accidentally deleting data you might need later. (We have certainly done that before!)

   Now you can save this altered shapefile as a copy without the hidden fields.

Export the altered shapefile

1. In the Contents pane, right-click the CountiesAge layer, and click Data > Export Features.

2. In the Export Features dialog box, for Output Location, browse to the Chapter08 folder. Name it **CountiesAge_Export**. Click OK.

   ArcGIS Pro adds the layer to your map.

3. In the Contents pane, open the attribute table for the new CountiesAge_Export layer.

4. Scroll to the right to confirm that the hidden fields aren't visible in the table.

   Wait! Maybe those fields are still in the file but hidden?

5. In the attribute table, click the Options menu, and click Fields View to confirm that the FUNCSTAT, ALAND, and AWATER fields are no longer in the list of fields.

   If you export a copy of a shapefile with fields hidden, those fields won't appear in the exported file.

6. Close the Fields View tab.

Work with multiple attribute tables

1. If necessary, open the CountiesAge_Export and CountiesAge tables.

    You can toggle between them by clicking their tabs. The tables can also be moved and docked elsewhere.

2. Drag the CountiesAge tab toward the center so that the docking tool (shaped like a plus sign) appears.

3. Hover over any of the four outside squares (displayed in tan and white) and release to split the view, with one table on each side.

4. To undo the split view, drag the table back to the center and "restack" it.

    Kind of cool, right? You may need to practice it again.

5. Save and close your project.

    You learned some handy things you can do with tabular data. You're a pro now! At least you have some new skills for managing the tabular attributes of your data layers. We hope you're feeling more confident in your ability to get around and manage basic tasks in ArcGIS Pro. Go take a break. We'll meet you at the next chapter!

USER STORY

Mapping COVID-19 testing sites

GISCorps is a volunteer organization operating under the auspices of the Urban and Regional Information Systems Association, or URISA. The organization conducts short-term mapping projects, emphasizing underserved communities, humanitarian aid, and disaster relief.

The organization's project to build the nation's most complete map of COVID-19 testing sites wasn't the first attempt. Coders Against COVID, an ad hoc group of computer experts and medical professionals, had already attempted to do so with its website findcovidtesting.com. But without a huge resource of volunteers, it stalled.

GISCorps had an army of GIS pros ready for action, but the logistics proved daunting. Ultimately, the two organizations connected their respective datasets through web services to aggregate efforts into a primary database both can use.

To read the full story, which appeared on *Esri Blog* in April 2020, visit links.esri.com/GISCorpsBlog. To explore the app itself, visit links.esri.com/GISCorpsSiteLocator.

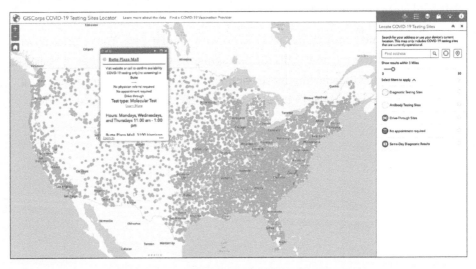

GISCorps map of COVID-19 testing sites shares data from Coders Against COVID.
Credit: GISCorps, Esri. Data source: GISCorps, Coders Against COVID, Esri, USGS, Garmin, FAO, NOAA, and EPA.

CHAPTER 9

Enriching your data

Data enrichment adds demographic or landscape information about the people and places that surround or are inside a data layer's locations. The results consist of a copy of your data layer with added attribute fields that can then be symbolized. That sounds a bit abstract, but you'll see in this chapter how powerful and useful this functionality is.

The Enrich tool references information in the ArcGIS Business Analyst™ dataset. Business Analyst incorporates several software apps (including Desktop, Enterprise, and Mobile) for ArcGIS users who work in demographic, business, spending, and census data industries, among others. Business Analyst requires additional licensing and consumes ArcGIS Online credits. But the license that ships with this book includes Business Analyst data and enough credits to give you a useful sense of its functionality. In this chapter, you'll limit the geography to minimize credit use. What are credits? We'll explain them in this chapter.

Although you may not always have access to Business Analyst in your GIS career, knowing about its data enrichment capabilities is important. Let's try it out.

Add and symbolize data

1. In ArcGIS Pro, start a new map project. Name it **Ch9_enrich**, and save it to your Chapter09 folder.

2. Click Add Data, and add the Ellicott_Tracts shapefile from the Chapter09 folder.

   These are census tracts for Ellicott City, near Baltimore, Maryland. You'll start by changing the symbology.

3. In the Contents pane, right-click the Ellicott_Tracts layer, and click Symbology.

4. In the Symbology pane, click the color square, and click the Properties tab.

5. For Color, click the down arrow, and click No Color.

6. For Outline Color, click the down arrow, and click Black.

7. Increase the Outline width to **2 pt.**, and click Apply. Save your project.

   Now the tract boundaries are displayed more distinctly. Ellicott City is the county seat of Howard County, Maryland, with a population of about 65,000.

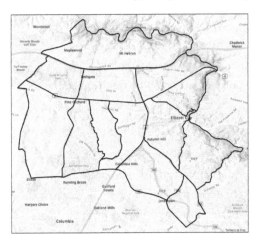

Enrich the tracts

You'll use the Enrich tool to add demographic data to the Ellicott City tracts.

1. On the Analysis tab, in the Tools group, click the down arrow to expand the tool gallery.

2. Scroll down, and click the Enrich tool to open it.

The Geoprocessing pane for the Enrich tool opens. The text at the top of the tool pane says that this tool consumes credits. That's good to know up front.

3. In the pane, for Input Features, click the down arrow, and click Ellicott_Tracts.

4. For Output Feature Class, keep Ellicott_Tracts_Enrich.

5. To the right of Variables, click the plus sign to open the Data Browser pane.

Wow, that's a lot of choices—Population, Income, Age, Households, Housing, Health, Education, and so on. Each of these categories contains multiple variables—making for far more choices than is immediately apparent. For example, you can search Spending > Food > Food at Home–Bakery Products and find data on rice purchases. Feel free to explore—and feel free to be a bit overwhelmed.

If the item of data you're looking for doesn't exist, you can probably use another item as a proxy. For example, spending on baby formula would be a good indicator of the relative number of babies in an area compared with another area. Yes, the data on babies is easy to find, but you get the idea of how, with such a huge library of data available, you can find a substitute for just about anything you want. Amaze your friends and family!

6. From the main categories list, double-click Housing to see the subcategories.

7. Double-click Owner & Renter to see the variables.

   Historical and projected data is available. When this book was being written, the most recent data was from 2022, but feel free to use more current data if you see it listed. Your results may vary from what you see in the book, but that's OK—the point is to explore the possibilities.

   For this item, you can select number or percentage. Many items also let you select an index comparing this data with a national or other type of average. Such index values can be useful.

8. Check the box for 2022 Owner Occupied HUs. (HU is short for housing unit.)

9. On the same line, check the percent (%) box to obtain a percentage, and uncheck the number (#) box to clear it.

   At the upper right, a count of the number of variables you've chosen appears in a black circle. In this case, just the one.

   > If you're interested in several variables, it's easy to acquire all of them at the same time. Just make sure you have the correct variables before using your credits.

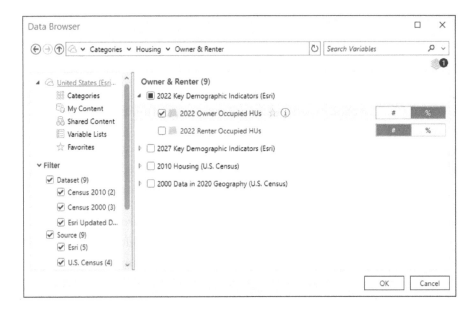

When you point to an item, a blue star appears, which lets you assign that item as a favorite for fast access. The blue *I* in a circle displays information on the item, including its source and a definition. If you aren't sure how an item is being counted, click the blue *I* to find out.

10. At the bottom of the pane, click OK.

11. At the top of the Geoprocessing pane for the Enrich tool, click Estimate Credits to check the potential credit use.

    The estimated credit consumption is less than one credit to enrich 13 tracts with one variable. Such a bargain!

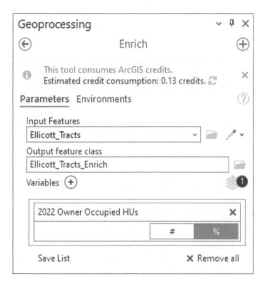

12. At the bottom of the pane, click Run.

    > When a geoprocessing tool runs, it may take several moments, so give it time.

    The map looks the same because the new layer was symbolized the same as your input layer. Let's see whether the table has any new goodies for us.

Review the table

1. In the Contents pane, open the attribute table for the Ellicott_Tracts_Enrich layer.

2. Scroll to the far right to display the enrichment field: 2022 Owner Occupied HUs Percent.

3. Close the table.

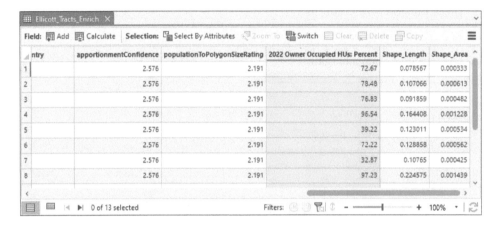

Symbolize the layer

You'll choose symbology for the new layer based on the 2022 Owner Occupied HUs Percent field.

1. In the Contents pane, right-click the Ellicott_Tracts_Enrich layer, and click Symbology.

2. For Primary Symbology, click the down arrow, and click Graduated Colors.

3. For Field, click the down arrow, and click 2022 Owner Occupied HUs Percent.

4. For Color Scheme, click the down arrow, and click the Oranges (Continuous) monochromatic scale.

   The map makes a couple of things clear. By area, housing in Ellicott City is predominantly owner occupied. In the Class values on the Symbology pane, three of the five classes have more than 74 percent owner-occupied housing. The map also reveals the locations of the predominantly renter-occupied housing. Those tracts, with lower owner-occupied percentages, are symbolized in light cream.

Chapter 9: Enriching your data

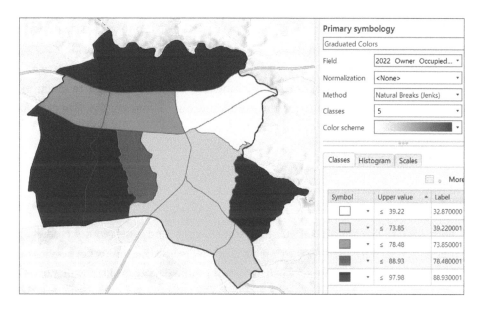

You're currently using the default Natural Breaks (Jenks) method to create these classes. Let's try a different method to explore the data a bit deeper.

5. In the Primary Symbology pane, for Method, click the down arrow, and click Equal Interval.

Do you notice the differences in the class values? The distribution didn't change at the top or bottom of the range, but there's a finer distinction among the classes in the middle.

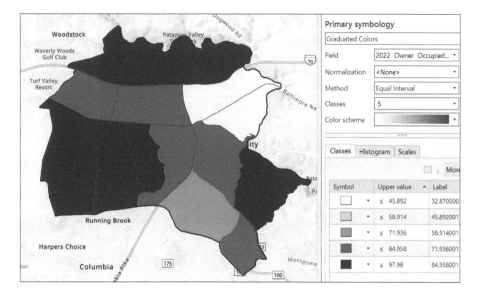

Pretty powerful stuff, right? Now that you've performed some data enrichment, reread the first paragraph in this chapter. It should make more sense after seeing how enrichment works. As you consider some of the available data you may be working with in the future, expect enrichment to be one of your go-to practices. Check out the next section to get the most value out of your credits. You now have another GIS skill to tuck away in your back pocket!

6. Save and close your project.

Credit use

In both ArcGIS Pro and ArcGIS Online, some actions and tools require credits. Generally, credits are used for storing data online, geocoding and other services, and accessing premium content such as demographic data from Esri. Still, most ArcGIS actions don't require credits, and when you do need them, you'll always receive an alert letting you know. When you need credits, you usually won't need many. For example, geocoding 125 addresses costs fewer than five credits.

At this point, you're probably wondering where you get credits or whether you have credits already. You received some credits through the license with this book. If you need to buy more credits, you can check the cost and learn more at links.esri.com/ArcGISCredits.

When you perform actions that consume credits in ArcGIS Pro, you'll always be given an alert: "You are using a tool that consumes ArcGIS credits. Click to estimate credits." If you click the words "estimate credits," you'll be able to decide whether you want to proceed. It's a little like the calorie counts on foods—are you sure you want to eat that? You want to make an informed decision.

Some options reduce your credit use. For data enrichment, you can use census tracts or even counties instead of block groups, choose a smaller area, or reduce the number of variables. For geocoding, be sure you keep your data in good shape to start with so that you don't need to repair it repeatedly. We have students who've geocoded a spreadsheet full of addresses again and again without thinking about it and found they had used all their credits. Just plan ahead, and you'll find that your credits can go a long way.

USER STORY

Tracking population growth with enriched maps

We can't say whether the US Census Bureau used the Enrich tool to make this map. But this is the type of map for which data enrichment can be powerful and useful. It uses an effective dichromatic color scheme (green to purple) to show the percentage change in population by county in the United States between 2020 and 2021.

Explore this map in greater detail at links.esri.com/CensusGrowingCounties.

US Census Bureau map of percentage change in population by US county 2020–2021.

Source: US Census Bureau, Vintage 2021 Population Estimates. Data source: US Census Bureau.

CHAPTER 10

Mapping x,y coordinate data

In a previous chapter, you learned how to map data that has an address using geocoding. Although geocoding is handy, what do you do if your data doesn't have an address? Think about a college campus: the buildings have addresses, but monuments, trees, and fountains don't. A water valve in the street, an animal den in the forest, and tent locations in a campground don't have addresses, either, but may need to be mapped. They all have a position on the earth that can be represented using latitude and longitude. We use latitude and longitude as our x,y coordinates. You'll learn how to use ArcGIS to map coordinate information.

In this chapter, you'll turn a table of trailhead locations with latitude and longitude coordinates into points on a map.

Chapter 10: Mapping x,y coordinate data

Add latitude and longitude data to ArcGIS Pro

1. In ArcGIS Pro, start a new map project. Name it **Ch10_trailheads**, and save it to your Chapter10 folder.

2. Click Add Data, and add the Trailheads.csv file in the Chapter10 folder.

   The file is added to the bottom of the Contents pane, under Stand-alone Tables.

3. In the Contents pane, right-click the Trailheads table, and click Open. Find the X field (indicating longitude) and the Y field (indicating latitude).

   You should be able to find those fields directly following the Name field, so you're good to go!

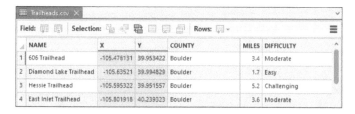

4. Close the table.

Convert coordinates into points on your map

1. In the Contents pane, right-click Trailheads, and click Display XY Data.

   The Display XY Data tool pane opens. ArcGIS Pro autocompletes the Input Table, Output Feature Class, X Field, and Y Field parameters.

   The Output Feature Class parameter—in this case, Trailheads_XYTableToPoint—saves the tool output as a feature class in your project geodatabase. You'll learn more about that way of storing spatial data in a later chapter.

   The GCS_WGS_1984 coordinate system is commonly used for worldwide data such as the basemaps in this project. But your data wasn't collected using that global coordinate system—it used something more specific.

2. In the tool pane, for Coordinate System, click the globe to the right of the text box, and browse to Geographic Coordinate System > North America > USA and Territories > NAD 1983. Click OK.

   The GCS_North_American_1983 coordinate system is specifically intended for North America.

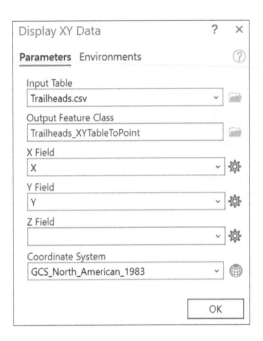

3. Click OK to run the tool.

When the tool finishes running, the new point layer is added to the Contents pane and the map. The map automatically zooms to the new point layer, as well. You now have a map of the trailheads in and around Boulder, Colorado. But before we hit one of those trails for a hike, let's do one other thing first and check out our options by way of ArcGIS Living Atlas of the World.

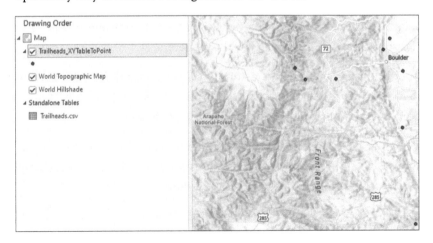

ArcGIS Living Atlas of the World

ArcGIS Living Atlas of the World is the foremost collection of global geographic information. Also known as ArcGIS Living Atlas, it contains data, maps, and apps that can be used in ArcGIS Pro or ArcGIS Online. Did we mention that most of it is free, and what isn't free is included in your ArcGIS Online subscription? ArcGIS Living Atlas items cover a lot of territory: everything from a 1796 map of Pittsburgh, Pennsylvania, to census block groups from the 2020 count—all 242,335 block groups! The categories include basemaps, imagery, boundaries, people, infrastructure, and the environment.

Searching through ArcGIS Living Atlas, you'll note that many items are listed as authoritative, so you can count on them being reliable, generally from the agency that created the data. Check back now and then to see what's new—there's always great content being added.

Browse ArcGIS Living Atlas at livingatlas.arcgis.com to see the variety and depth of information that's available. We find it handy to use the site to locate what we want and then tell ArcGIS Pro what we want, as you'll do in this chapter. You can search by various criteria and find ready-to-use layers you probably didn't even imagine existed. And don't forget the part about it being free.

Add data from ArcGIS Living Atlas to your map

You've heard that this area of Colorado can be prone to flooding. So before taking any hikes, you'll add data to help you determine whether it's safe to hike in this area. You'll search ArcGIS Living Atlas and include some of its content in your map. You can reach ArcGIS Living Atlas in several ways in ArcGIS Pro, but for now you'll use your good friend, the Add Data button.

1. Click Add Data.
2. In the left pane, expand Portal, and click Living Atlas.

Top 20 Essential Skills for ArcGIS Pro

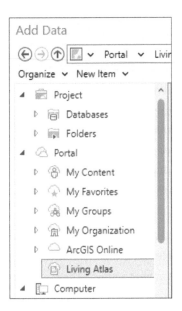

3. In the Search Living Atlas box in the upper right, type **live stream**, and press Enter.

   Many choices appear. The title describes what the data layer contains.

4. In the list of titles, locate Live Stream Gauges.

   Next to the title are several blue icons. You can point to each icon to learn what it does in a ScreenTip. The icons adjacent to the Live Stream Gauges layer are, from left, "Authoritative: Recommended by Esri," "Living Atlas: Esri-curated content," and "Sharing level: Public." The Type field indicates the type of data layer (for example, feature layer or map image layer). The Date Modified field is self-explanatory. Owner tells you the person or organization that created or contributed the data layer. Path is the URL for the layer.

5. Click Live Stream Gauges to add it to the Name box at the bottom.

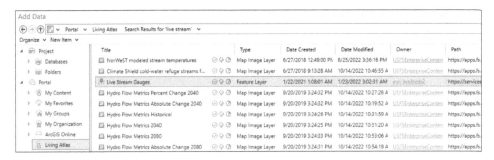

6. Click OK to add the layer to the map.

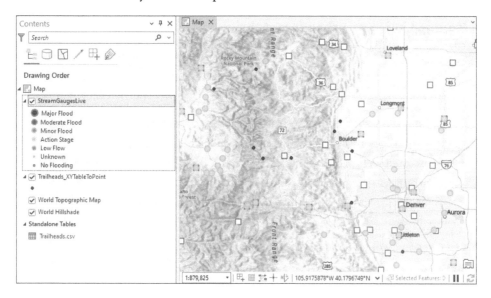

Keep in mind that this layer monitors *live* stream gauges. The map shown in your book is a snapshot from a point in the past, when this lesson was written. As you work through these tutorial steps, you'll be viewing the layer at another time, and it may not look the same. In the map image, most of the map is No Flooding (white squares) and Unknown (blue circles), which bodes well for our hike. Is it safe for you to go hiking?

7. If it's safe to hike, save and close your project, and hit the trails!

You've learned both ways of converting tabular data into point features on a map: geocoding and displaying x,y data. You've even learned about the data treasure trove in ArcGIS Living Atlas. You're becoming a serious GIS professional!

In this chapter, you started with data that had already been collected and was in a usable format, but how would you collect data that doesn't have an address? A common way is with a smartphone. Employees in the field may use a phone to add project data; the GPS in the phone adds the location coordinates. For more precise work, surveying equipment provides amazing accuracy. Prices for such equipment range widely (some robotic equipment may cost thousands of dollars). For most work, your smartphone provides enough accuracy. After all, if you want to find a tree, if the GPS functionality in your phone puts you within a couple of feet, that's close enough. For surveying a parcel, though, not so much.

USER STORY

Mapping helps an urban forest thrive

Trees typically don't have an address, but they can be mapped using their location coordinates. Check out how the University of North Alabama (UNA) uses GIS data to track trees in its urban forest.

By tracking trees with GIS, UNA knows which trees are likely to thrive in certain spots or soil types and which will restore needed biodiversity. Urban forests, when designed well, help keep water from flooding or pooling at the surface, which can affect the soil, compromise plant growth, and cause erosion. Also, urban forests play a role in mitigating global warming by capturing and storing carbon dioxide.

So far, UNA has mapped 79 plant species, 1,487 trees, and 42.79 acres of turf that make up the university's 147-acre campus. It's all part of the UNA Arboriculture Management field map, which is used to address the variety of challenges that come with managing urban forests, such as removing diseased or risk-laden trees, diversifying the tree population in planning, and scheduling routine maintenance.

To learn more about the UNA project, consult "How Mapping Trees Helped a University's Urban Forest Thrive" on *Esri Blog*, on February 10, 2022, at links.esri.com/UNAForest.

University of North Alabama Arboriculture Management field map helps users monitor data on campus flora.

Credit: University of North Alabama. Data sources: University of North Alabama, Esri, HERE, Garmin, INCREMENT P, USGS, EPA, USDA.

CHAPTER 11

Editing feature data

In this chapter, you're encouraged to let your creativity run wild! We've found that many of our students have fun editing state boundaries, and that's what you'll be doing. We hope you have fun, too. Yes, it's an odd thing to have fun with, but you probably will, and we're glad.

You aren't likely to need to edit state boundaries, because they don't change much, but you'll be using numerous datasets you'll want to edit in your work. Suppose you were working on habitat restoration for a species—the range of a species may change. If you work for a business with salespeople in the field, you may need to adjust the territories each salesperson covers. Working for a city, you may need to change the district boundaries of your elected officials. For every industry, there are countless examples like these.

The Undo button—which you used earlier—is probably even more important in feature editing. Oops, I moved that edge too far. Let me go back. Uh-oh, I accidentally moved a vertex (a point where two or more edges meet) and made a big mess. When you can, use the Undo button instead of trying to manually fix a mistake you just made. Thanks, Undo!

Add a shapefile

1. In ArcGIS Pro, start a new map project. Name it **Ch11_editing**, and save it to your Chapter11 folder.

2. Click Add Data, and add the States.shp file from the Chapter11 folder.

3. Zoom in to a US state of your choice so that you can see the state boundary. (For this project, we're using California as the example).

    This is the continental United States only. Sorry, Alaska and Hawaii!

Edit a polygon's shape

1. On the View tab, in the Navigation group, click Navigator.

    > **Hint:** You'll find the navigator at the lower left of your map. Click the up arrow to get full navigational control. Navigating the map while editing features can be tricky, so using this control helps.

2. On the Edit tab, in the Selection group, click Select to activate the Select tool.

3. Move the pointer over the map, and click once anywhere in California.

    The state is highlighted to indicate that it's been selected.

    We've modified our settings so that our selected features are red, which makes them easier to see. You can change the color; otherwise, selections are cyan by default.

4. On the Edit tab, in the Tools group, click the Edit Vertices tool.

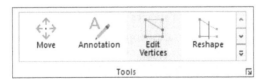

Your state boundary changes to tiny green points called vertices. These points are connected by lines, which make up the feature. A toolbar is added to the bottom of the map frame, and the Modify Features pane opens to the right. Now you're ready to have a little fun!

5. Zoom in to a smaller section of your state boundary so you can see the individual vertex points as little green squares.

   If you need to pan, try using the navigator. You can also use the Explore tool on the Map tab. Doing so reverts to the selection outline (in cyan). When you have the location you want, return to the Edit tab and click the Edit Vertices tool to display the vertices.

6. Click any one of the vertices (green squares), and drag it to a different location to begin reshaping your state.

7. Move more of the vertices to get the feel of reshaping the state boundary.

8. When you've finished, click anywhere outside the state boundary to complete the edits.

   Here's our wildly reshaped California:

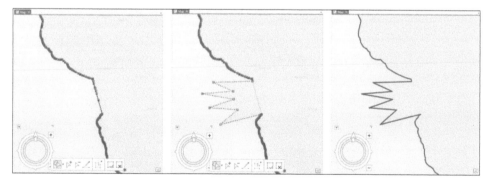

We got a little silly with this example and exaggerated the shape, but now you understand how editing can be helpful for, say, reshaping a city boundary to include a newly annexed area or for something as simple as a change in the fence line for a work site.

Move a polygon

1. Zoom out so you can see your whole state. Make sure the state is selected (outlined in cyan) so that the editing tools will affect only that state.

2. On the Edit tab, in the Tools group, click the Move tool.

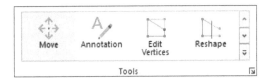

The Move tool adds a dashed black-and-white outline to the state's perimeter. It also changes the pointer to a four-headed arrow.

3. Click anywhere in your state, hold the left mouse button down, and drag the state to a different location. Release the mouse button.

4. Click anywhere outside your state to complete the edit.

Our California is now an island in the Pacific Ocean, as early mapmakers believed. Search for "Island of California" if you want to learn about one of the most famous errors in cartography, from the 17th century.

Split a polygon

This editing tool is used to split polygon features.

1. On the Edit tab, in the Features group, click Modify.

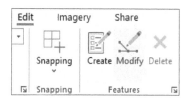

The Modify Features tool pane appears. Check out all the editing choices! We've collapsed the other categories to focus on the tool we want to use—the Split tool.

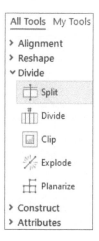

2. In the Modify Features pane, expand Divide, and click the Split tool.

3. Click anywhere in your state.

   A new toolbar is displayed at the bottom of the map.

   The Line tool is activated by default. That's the tool you want. Thank you, ArcGIS Pro!

4. With the Line tool activated, click one side of your state, and drag to the other side, extending beyond the state boundary. Don't release just yet.

   A dashed line appears between the initial drag point and the potential next click, as illustrated in the image.

5. Double-click to split your state.

   ArcGIS Pro has divided your state into two separate polygons, both of which are still selected (in cyan by default).

6. If you're happy with your work, on the Modify Features toolbar, click the Finish button (the square with the green check mark).

7. Close the Modify Features pane.

Move a split polygon

You can have a bit more fun with the Move tool. Try this.

1. On the Edit tab, in the Selection group, click Select.

2. Click one of the polygons in your split state.

3. On the Edit tab, in the Tools group, click Move.

   The Modify Features pane gives three options: Move, Rotate, and Scale. Let's try all three!

   Move is activated by default, so you'll move one of your polygons first.

4. Click your selected polygon, and drag it to another location.

   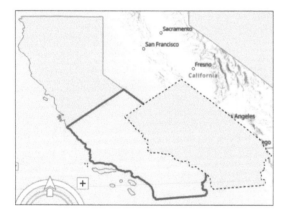

   The map shows both the original location (solid outline) and the new location (dashed outline) for the polygon.

5. When you're happy with the location, on the Modify Features toolbar, click Finish.

Rotate a polygon

1. With your polygon still selected, click Rotate in the Modify Features pane.

   A green rotation wheel appears on your polygon.

2. Click the green rotation wheel, and drag in a circular direction until your polygon is rotated the way you want.

3. On the Modify Features toolbar, click Finish.

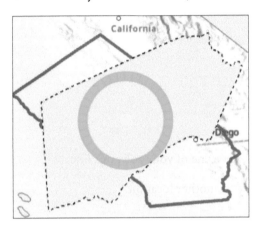

Scale a polygon

1. With your polygon still selected, click Scale in the Modify Features pane.

   A box with grab handles appears around your polygon. We're from Southern California (which, of course, we think is best!), so we'll scale Southern California to be even larger.

2. Click one of the grab handles, and drag it outward until your polygon is scaled the way you want it.

3. When you're satisfied, click Finish.

4. On the Edit tab, click Save to save your playful edits.

5. If you need to confirm, click Yes.

6. Save and close your project.

   Hope you had as much fun with that as we did! In the midst of all that fun, you learned some valuable editing skills that you'll be able to apply to future work projects. See how we snuck that in there? You're welcome.

The issue of accuracy

We all want our data to be perfect. We want the location to be precise and the lines to be placed in just the right place. That's a fine goal but not always a good idea. We'd better explain. Suppose a regional agency needed land-use data to analyze potential housing construction. The agency collected the information by city block because that was as accurate as the analysis required. Cities got the data from the agency and found it useless for their needs because they work at the parcel level—more detailed than city blocks. Had the data been gathered at the parcel level, the regional agency and the cities could both have used it.

Let's say you're making a map of water lines. The planning department needs to know whether there is water service to a particular building, whereas the water department needs greater accuracy and detail to maintain the water system. For the planners, the data can be correct to within several feet, but the water department needs even greater accuracy to safely dig in the street and lay a pipeline.

Of course, part of the power of GIS is that everyone can share data. The accurate data from the water department can serve everyone who uses that data. If the accuracy were only to the level needed by the planning department, it would be of no help to other users. We don't want multiple departments having to maintain a layer for pipelines. The job of maintaining that layer goes to the people who create the data, which is the water department, in this case. The key to making that data valuable to many applications is to make it as accurate as the most demanding user of it needs it to be.

However, don't get carried away with making your data as accurate as possible. If you're mapping the potential track of a storm, you wouldn't need more small-scale accuracy than if you were mapping tree locations. Being more accurate may seem better, but it isn't always useful and can be a poor use of resources.

USER STORY

Spreading the advantages of broadband

Giving every US resident opportunities for economic mobility, jobs, education, services, and more requires equitable access to broadband.

In 2016, the City of Dubuque, Iowa, introduced the Dubuque Broadband Acceleration Initiative, an effort to focus on public-private collaborations that reduce the cost and time required for broadband expansion. The city used location intelligence from a GIS to support the efforts of residents to work with providers, secure funding, and fast-track local broadband investment.

Using ArcGIS Survey123, they set up the Dubuque Broadband Services Survey to collect baseline data on current use, speed, and bandwidth, which clarified the need for services from households across the city.

Survey results (refer to the image) clearly showed areas of the city where people are disproportionately disadvantaged. For example, 30 percent of survey recipients said they requested faster internet speed and were told it wasn't available. Using this information, the city determined priority areas and developed a long-term plan to apply for grant funding.

To read more about the Dubuque Broadband Acceleration Initiative, visit the Esri case study "How Dubuque Is Using GIS to Make the Case for Broadband Funding" at links.esri.com/DubuqueBroadband.

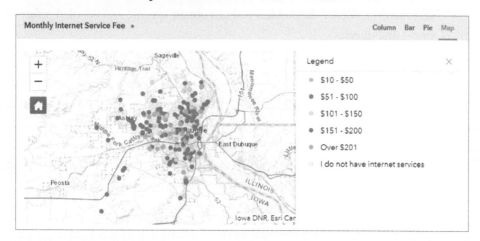

Monthly Internet Service Fee map for the Dubuque Broadband Acceleration Initiative.

All content © 2006-2023 City of Dubuque and its representatives. All rights reserved. Map data source: City of Dubuque, Iowa, Iowa DNR, Esri Canada.

CHAPTER 12

Performing data queries

In this chapter, you'll learn how to query data, or ask questions, and use the answers. That may not sound exciting, but it is when you get the hang of it. Querying data is another powerful technology in GIS. Depending on the information in your data tables, you can obtain specific and detailed information using queries.

Attribute queries are based on the tabular information of your data. Suppose you had a spreadsheet of apartments that included rents and amenities. You could find apartments with a rent in a particular range that also had covered parking. You could then see where those apartments are and use spatial queries to find which of those apartments are located within a particular neighborhood. In this lesson, you'll use the census data you've been working with to find areas with a high proportion of seniors.

This chapter is short but powerful. Get to it!

Add a shapefile

1. In ArcGIS Pro, start a new map project. Name it **Ch12_queries**, and save it to your Chapter12 folder.

2. Click Add Data to add CountiesAge.shp from your Chapter04 folder.

   If you can't find the shapefile in the Chapter04 folder, grab a copy of the CountiesAge layer for California in the Chapter12 folder.

   Remember when you created this shapefile by joining the census table to your state's counties polygons?

   This exercise is shown with California counties as the example, but feel free to work with your own state of choice from chapter 4 to explore its senior population.

Write an attribute query

1. On the Map tab, in the Selection group, click Select by Attributes to open the Select by Attributes dialog box.

Querying the correct layer's attributes

Right now, you have only one layer of data in your map, so by default it's the active layer (cyan highlight in the Contents pane). The Select by Attributes tool populates the input. But if you have more than one data layer in your map, you can manually populate the tool input in other ways.

One, you can click the correct layer in the Contents pane to highlight it and then click Select by Attributes. Two, you can choose the correct layer from the drop-down options for those inputs.

Input Rows is autocompleted with CountiesAge.

2. You want to create a new selection, so for Selection Type, keep New Selection.

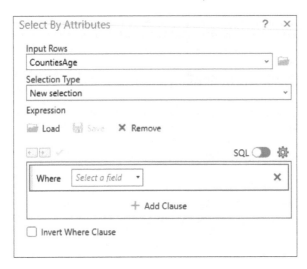

It's time to write your first query.

3. Under Expression, for Where, click the down arrow, and click Seniors.

    This change adds two more text boxes: Operator and Value.

4. For Operator, click the down arrow, and click Is Greater Than.

5. For Value, type **15**.

You're starting with a simple query to select all counties where the senior population is greater than 15 percent.

6. If your query is correct, click Apply.

    The dialog box stays open, allowing you to review the query results on the map. The selected counties are outlined in the map. The query selected many of the counties in the state, so you'll modify the selection.

7. In the Select by Attributes dialog box, change the value of 15 to **20**, and click Apply.

    > **Hint:** In step 6, if you clicked OK instead of Apply, click Select by Attributes again to re-create the query. It's good practice!

What a difference! In the first image, the map shows California counties with a senior population of greater than 15 percent. In the second image, the map shows counties with a senior population of greater than 20 percent.

8. Click OK to close the dialog box.

Examine the selection totals

So how many counties did your query show a result for? As is typical with ArcGIS Pro, you can find the answer in several ways—but here's the quickest way.

1. Examine the lower right of the map, and you'll notice Selected Features: 18.

   There you go! But suppose you want to know which layers have a selected feature?

2. In the Contents pane, click List by Selection.

   Eighteen counties are selected in the CountiesAge layer. This view is especially helpful when you have selections on several different data layers.

This number indicates the total number of counties selected. But you still want to know which counties contain a selection. You'll have to try something else to get that information.

3. In the Contents pane, click List by Drawing Order, right-click the CountiesAge layer, and click Attribute Table.

   The attribute table appears, with the selected counties highlighted. Now you can determine exactly which counties have a senior population of more than 20 percent. The attribute table also indicates that 18 of 58 counties are selected.

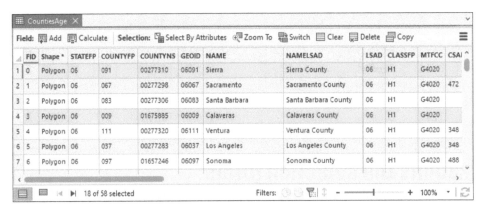

4. In the lower left, next to the number of selected features, click Show Selected Records (the button directly to the right of Show All Records).

   Now the table displays only the selected counties—the counties you're interested in. This view doesn't permanently affect your data, by the way, so don't worry.

5. Click Show All Records to see the complete set of records in the table, and then close the table.

Create a shapefile of the selected features

1. In the Contents pane, right-click CountiesAge, and click Data > Export Features.

   The Export Features tool pane's Input Features box is autocompleted with CountiesAge.

2. For Output Feature Class, browse to the Chapter12 folder, and name the shapefile **CaliforniaSeniors** (remember not to use spaces in the file name).

Chapter 12: Performing data queries

3. Click OK to export the selected counties.

   ArcGIS Pro will add the exported file as a new layer to the Contents pane.

Clear the query

You can clear, or erase, your query results in multiple ways. Here's the simplest first.

1. On the Map tab, in the Selection group, click Clear.

   All selections you made earlier are cleared, which may not be what you want. If you have multiple layers with multiple selections, you can clear the selection in just an individual layer.

2. In the Contents pane, right-click the CountiesAge layer, and click Selection.

   In the options for Selection, Clear Selection is unavailable (light gray). You already cleared all selected features for the whole project. But it's still useful to know how to clear a selection in an individual layer.

3. Save and close your project.

   Congratulations! You've learned the basics of attribute queries. This skill will come in handy a lot more often than you may think. In your work, you'll find that performing a query with multiple criteria is not only valuable but it's not difficult either. In the next chapter, you'll learn how to make selections based on the location of features rather than their attributes. Movin' on!

USER STORY

Site suitability helps states move to renewable energy

The state of Kentucky is transforming from a coal powerhouse to a locus for renewable energy generation. Once the leading producer of coal in the United States and still one of the top three coal-producing states, Kentucky is nonetheless planning a future powered by alternative energy sources—hydropower, biomass, and solar.

The Kentucky Energy and Environment Cabinet used GIS to conduct site suitability analysis on land parcels available for development. This analysis evaluated sites on the basis of construction criteria and then gave each parcel a score. These activities addressed the concerns of solar plant providers—favorable slope, land classification (barren land, mixed forest, cultivated crops), access to electric transmission lines, population density, proximity to the habitats of threatened species, and status as federal or protected lands.

To read more about the Solar Site Potential in Kentucky project, go to "Finding a Home for Solar: Kentucky Maps Prime Renewable Energy Sites" on *Esri Blog*, July 22, 2021, at links.esri.com/KentuckyEnergySites.

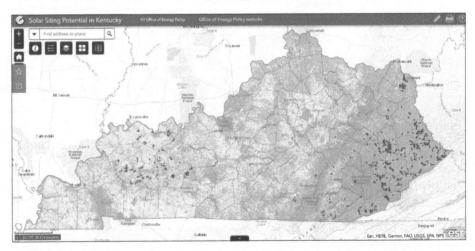

Kentucky web app lets solar developers know which sites are best for building solar plants.

Map data source: Kentucky Energy and Environment Cabinet, State of Kentucky, Esri, HERE, Garmin, FAO, USGS, EPA.

CHAPTER 13

Performing location queries

In this chapter, you'll learn how to query data to create selection sets based on the location of features in relation to other features. Location queries are also known as spatial queries. Think of it this way: spatial queries can select something based on where it is. They may not sound all that exciting, but they're undeniably useful and can be performed only in GIS.

A spatial query can reveal all sorts of necessary information: the homes within 500 feet of a hazardous waste spill, parcels that a creek crosses or touches, communities with a park in them, or sales territories that overlap. It can tell you how many schools are located in the ash plume of a wildfire. It can tell you which industrial parcels are within 1,000 feet of residential parcels. These aren't trivial matters—they're immediately useful for dealing with a disaster or creating the language for a new city ordinance. ArcGIS Pro provides many ways to select features spatially using categories to do so: for example, within a certain distance, touching, contains, intersect, within, and identical.

That's enough telling. It's time for showing. Let's jump into some spatial queries.

Add shapefiles

1. In ArcGIS Pro, start a new map project. Name it **Ch13_spatial**, and save it to your Chapter13 folder.

2. Click Add Data to add tl_2022_06_place.shp from your Chapter01 folder.

   You downloaded this file of cities back when you were an ArcGIS newbie in chapter 1.

   > *If you're using your own state, you'll find the place shapefile in your Chapter01 folder, but it will have a different number in its name—that's the file you want to add. You can find replacement data for California in C:\GIS20\Chapter13\ReplacementData.*

3. Add CaliforniaSeniors.shp from your Chapter12 folder.

   > *If you're using a different state, you should have made a CountiesAge shapefile for your own state in chapter 4 and then created your own Seniors shapefile in chapter 12. If that's the case, you want to use that file from your Chapter12 folder.*

   CaliforniaSeniors is added to the Contents pane, listed above the tl_2022_06_place layer. In other words, it draws last and ends up obscuring the layer below it. Let's fix that.

4. In the Contents pane, drag CaliforniaSeniors below the tl_2022_06_place layer.

   That looks better!

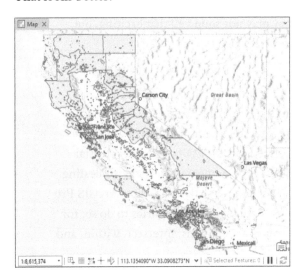

Write a location query

1. On the Map tab, in the Selection group, click Select by Location to open the Select by Location tool pane.

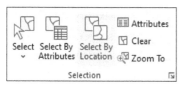

2. For Input Features, click the down arrow, and click tl_2022_06_place.

3. For Relationship, click Have Their Center In.

4. For Selecting Features, click CaliforniaSeniors.

   You want to create a new selection.

5. For Selection Type, keep New Selection.

So, what do all these parameters mean? You're asking ArcGIS Pro to select the places (Input Features) that have their center in the counties layer of high senior populations (Selecting Features). Before you click OK, think about what the results should look like. You should generally see the places that are on top of the counties layer being selected.

6. Click Apply.

   If the results don't look like what you expected, revise the options until you're asking the right question in your query. Think of your query as a question: it's good practice when you create queries. It's too easy to get it wrong on the first try. The good news about examining your question before running the query is that you haven't committed to anything yet. Just redo the process until you get the result you expect.

In this case, your selection resembles the image (though your colors may vary, of course). Only the cities with their centers within the counties layer have been selected. Perhaps these cities can be targeted for funding or social services for their senior population. GIS can be a handy tool for good!

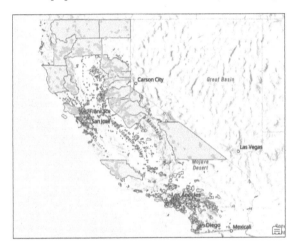

7. If the selection looks correct, click OK.

   Suppose you want to estimate how much funding all these senior programs in the selected cities require. You'd need to check the number of cities that your query identified. For a little practice, let's check the total in two ways.

8. In the lower right of the map, look for the number of Selected Features. Handy!

9. Alternatively, open the attribute table for the tl_2022_06_place layer, and scroll through the list of city names. Look for the number of selected and total records at the lower left of the table.

Export a table for the selected features

Perhaps someone else in your organization wants to build a spreadsheet to track outreach efforts for those selected cities. No problem—you can help with that!

1. In the Contents pane, right-click the tl_2022_06_place layer, and click Data > Export Table.

Chapter 13: Performing location queries

In the Export Table tool pane, the Input Table autocompletes correctly with the places layer.

2. For Output Table, save the file to your Chapter13 folder, and name it **PlacesInHighSeniorCounties**.

   ArcGIS Pro automatically adds the .dbf file name extension. This database format can be opened in Microsoft Excel and saved in any format.

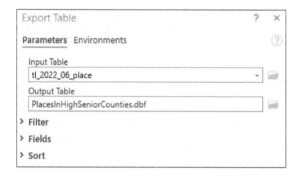

3. Click OK to save the table with information for only the selected cities.

   The tool runs quickly, but you can briefly see a progress bar. If you blink, you may miss it. After the export finishes and the progress bar disappears, nothing else happens in ArcGIS Pro. You may wonder, "Did it work?" Yep!

4. In File Explorer, browse to your Chapter13 folder to confirm that you see the new .dbf file.

5. In ArcGIS Pro, clear the selected features in your project.

   Clearing any selection after you've finished an operation is a good habit to adopt.

6. Save and close your project.

   Congratulations! You've learned how to perform location queries of your data. You can't even imagine how useful this skill will be. Now that you know how to perform attribute and location queries, you're on your way to becoming a query master! Do you feel the power?

USER STORY

Relying on GIS to help manage park resources

Challenge: Missouri's state park system contains 91 state parks and historic sites spread over 160,000 acres. With more than 2,000 structures, 3,500 campsites, 194 cabins, nearly 2,000 picnic sites, and more than 1,000 miles of trails, Missouri State Parks needed to develop a way to manage and maintain all these various assets.

Solution: The organization established an enterprise GIS to inventory and manage thousands of critical assets across the park system. This central GIS was transformative for the park system because it previously had no electronic record of assets or an effective way to manage the assets at the park, regional, or state level. This real-time information is displayed in a dashboard, allowing employees to make better resource management decisions.

To learn more about the Missouri State Parks system for tracking park assets, visit https://mostateparks.com.

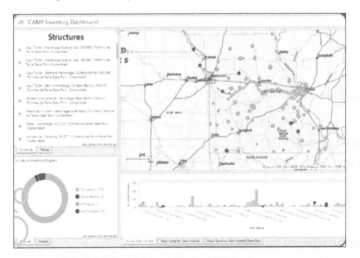

The Missouri State Parks Central Asset Management Program (CAMP) dashboard keeps track of the state's 91 state parks and historic sites spread over 160,000 acres.

Source: GovLoop, Esri, Missouri State Parks, Missouri Department of Natural Resources. Map data: Missouri Department of Natural Resources, Esri, HERE.

CHAPTER 14

Using geoprocessing tools

Many of the questions you expect GIS to answer are handled by geoprocessing tools, the cornerstone of spatial analysis. The most used tool may be the Buffer tool, which can find all the students living within a mile of a school or tell you which parcels along a stream have special development rules because of their proximity to the water. We dare you to perform those tasks for a large area without using GIS. You'll also learn four other handy tools. Once you familiarize yourself with how they work, you'll understand the flow for many other geoprocessing tools.

In this chapter, you'll learn how to use powerful geoprocessing tools to analyze data layers in ArcGIS Pro. Our favorite is the Visualize Space Time Cube in 3D tool, but that's just because it has a great name. We'll introduce you to some of the many practical tools, even if you aren't trying to visualize a space-time cube.

You can access the various geoprocessing tools in several ways. In this chapter, you'll use all the different methods as you explore the five most frequently used tools. Let's start with the most advanced version of the Buffer tool: Pairwise Buffer.

> *"Pairwise" indicates the version of a tool that generally executes the operation faster and sometimes offers more functionality. For our purposes, there won't be much difference, but we'll use the instructions for using the Pairwise tools.*

Pairwise Buffer tool

The Pairwise Buffer tool creates buffer polygons around input features to a specified distance. It's used to show proximity to certain features and can be used with point, line, or polygon features (see examples of each in the image).

> *You'll learn more about the dissolve option for buffers later in the chapter.*

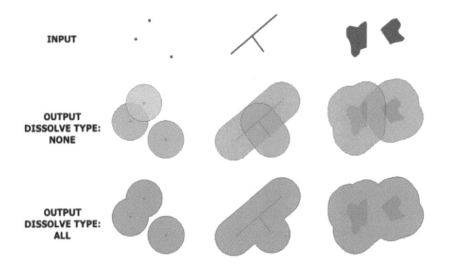

Create a buffer

1. In ArcGIS Pro, start a new map project. Name it **Ch14_geoprocess**, and save it to your Chapter14 folder.

2. Click Add Data, and add GroceryStores.shp from the Chapter14 folder.

   The shapefile consists of a sample of grocery stores in the Lincoln, Nebraska, area.

3. On the Analysis tab, in the Geoprocessing group, click Tools.

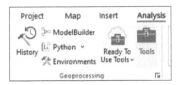

The Geoprocessing pane opens to the right. It opens with predetermined favorites, but that list is just the tip of the geoprocessing iceberg.

4. Click Toolboxes to view all the toolboxes.

   Inside the toolboxes are toolsets that contain the tools. Depending on your license, different tools are available for you to use.

   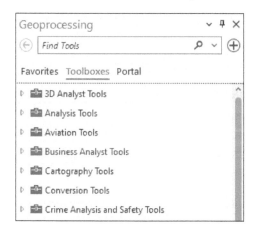

5. Expand the Analysis Tools toolbox, and expand the Pairwise Overlay toolset to view the Pairwise Buffer tool.

   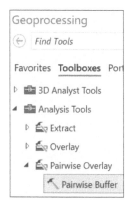

6. Click Pairwise Buffer to open the tool.

   As noted earlier, Pairwise Buffer is an improved version of the standard Buffer tool.

7. In the tool pane, click the pin in the upper right to keep the pane open in your workspace.

8. In the Contents pane, click the GroceryStores layer, and drag it into the Input Features text box in the tool pane.

> **Hint:** Did you know that you can drag any layer from the Contents pane into any of the tools? Pretty cool, right?

ArcGIS Pro autocompletes the Output Feature Class and chooses a location.

9. For Output Feature Class, click the browse button (the folder icon), and save the file to your Chapter14 folder with the name **GroceryStores_Buffer**.

You want to see what is within a half mile of GroceryStores.

10. For Distance, type **0.5**.

11. Adjacent to the distance value you just set, click the down arrow for Unknown, and click US Survey Miles.

12. Keep all other default parameters, and click Run at the bottom of the pane to run the Pairwise Buffer tool.

A progress bar at the bottom of the pane tracks the progress, and a pop-up summary of the analysis appears. When the tool has finished running, the buffers are added to the map and the Contents pane.

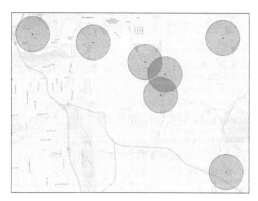

Change the symbology of the buffers

Those little buffers sure are cute! Your buffers may be drawn with a different color, but that's okay.

Generally, buffers are more useful on a map if you can see the features beneath them.

1. In the Contents pane, right-click the GroceryStores_Buffer layer, and click Symbology.

2. In the Symbology pane, click the blue square next to Symbol.

    You want the Properties tab to be active. If your Properties tab opened, great.

3. If your Gallery tab is open, click Properties.

4. Under Appearance, for Color, click the square to the right, and click No Color.

5. At the bottom of the pane, click Apply to remove the color fill for the buffers.

    For Outline, the color is unavailable. That may seem odd, but you can get to that another way.

6. Under Properties, click the Layers icon.

7. Under Appearance, change the Color square to Mars Red.

> **Hint:** Remember that if you point to the colors in the palette, you'll see their names.

8. Change the Outline Width to **3 pt**.
9. At the bottom of the Symbology pane, click Apply to change the look of the buffer outlines.

Now you can see other features underneath the buffers, such as streets, parks, and lakes. That's helpful because it gives your grocery store points some context.

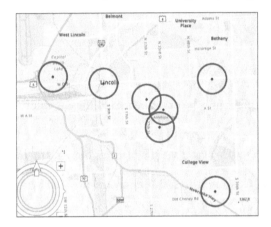

To dissolve or not to dissolve?

So, what's the difference in the dissolve options for the Buffer tool? The default is to use no dissolve. With no dissolve, each input feature is surrounded by its own buffer polygon. If the buffer polygons don't overlap, there's no difference between dissolve or no dissolve because boundaries are distinct. You'd notice the presence or absence of dissolve only if the buffers overlap, as they do in the first image. It looks a bit like a Venn diagram. And if many buffers overlap, it can look like a mess—especially if the buffers are symbolized with outlines and no fill. In that case, the buffers can look like bad Olympic rings!

The alternative is to dissolve all output features into a single feature, which results in only one larger buffer polygon for any buffers that overlap, as indicated in the second image. The overall look is a little cleaner in this example. If your map contains several overlapping buffers, dissolving helps your map present a clearer message. Neither choice is the right answer all the time. Your decision should vary, depending on context and the purpose of your map.

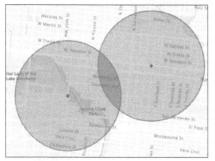 

No dissolve applied to buffers (*left*). Both output buffers dissolved into a single feature (*right*).

Measure the buffer distance

How do you know that the buffers have a 0.5-mile radius? Let's measure them to verify.

1. Zoom in to one of the buffers.

2. On the Map tab, in the Inquiry group, click Measure.

3. If you clicked the Measure down arrow, click Measure Distance from the options. Otherwise, the Measure Distance tool opens by default in the upper left of the map.

4. To see other measuring choices, in the Measure Distance dialog box, click the down arrow for the ruler.

    The measurement categories include Measure Distance, Measure Area, and Measure Features. For now, you'll stick with measuring distance.

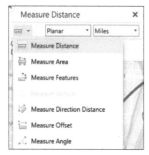

5. In the Measure Distance dialog box, keep the parameter for Planar, but choose Miles for the distance unit, because you used miles for your buffers.

   Notice that your pointer has turned into a measuring tool.

6. Click a grocery store point in the middle of a buffer boundary, and double-click anywhere on the buffer line.

   The Measure Distance tool verifies that your buffers are indeed a half mile around each grocery store.

7. Close the Measure Distance tool.

   If you ever need a quick measurement of distance or area, remember that Measure tool. Back pocket!

Merge tool

The Merge tool combines multiple input datasets into a single new output dataset. This tool can combine point-to-point, line-to-line, or polygon-to-polygon feature classes or tables. The image shows a polygon example.

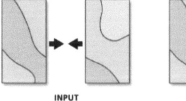
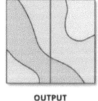

INPUT   OUTPUT

Create a merge

1. In the Contents pane, uncheck the GroceryStores layer and the GroceryStores_Buffer layer to turn them off. Collapse the layers to simplify your view of the Contents pane.

2. Click Add Data, and add the Southwest and Pacific_NW layers from the Chapter14 folder.

3. In the Contents pane, press and hold the Shift key, and click both the Southwest and the Pacific_NW layers, right-click, and click Zoom to Layer.

   The map zooms to the extent for both layers.

4. On the Analysis tab, in the Tools group, click the down arrow to expand the Tools gallery.

5. Scroll down in the gallery to the Manage Data group, and click Merge to open the tool in the Geoprocessing pane.

6. For Input Datasets, choose Southwest and Pacific_NW (one layer in each text box).

   Alternatively, you can drag the layers from the Contents pane.

7. For Output Dataset, browse to your Chapter14 folder, and name the file **SW_PNW_Merge**.

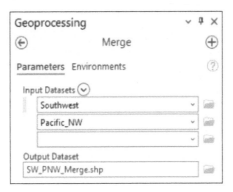

8. Click Run to run the tool.

    The tool generates a progress bar and a summary, just as it did with the Buffer tool. All the geoprocessing tools provide this information.

    When the tool has finished, the output layer is added to the map. Depending on the default colors, you may not be able to see that the new layer contains six US states (for Southwest and Pacific Northwest).

9. In the Contents pane, turn off the Southwest and Pacific_NW layers to better see the new merged layer.

10. Right-click the new layer to open its attribute table, in which you can confirm all six states. Close the attribute table.

Append tool

Whereas Merge is used to combine multiple feature layers into a new feature layer, the Append tool is used to add features to an existing feature layer. This image illustrates an append with line features, but you can use your Southwest and Pacific_NW polygon layers to see how an append operation works.

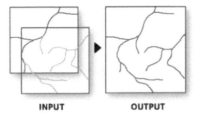

Append features to an existing feature layer

1. In the Contents pane, turn off the SW_PNW_Merge layer, and turn on the Southwest and Pacific_NW layers.

2. In the Contents pane, drag the Southwest layer above Pacific_NW.

3. On the Analysis tab, in the Tools group, search the Tools gallery for the Append tool, and click it to open it in the Geoprocessing pane.

   Immediately, the tool pane provides a helpful reminder: "This tool modifies the Target Dataset." That's what we want to do, so we're on the right track.

4. For Input Datasets, add the Pacific_NW layer.

5. For Target Dataset, add the Southwest layer.

   The tool provides options to narrow the append using an expression and to specify field matching. It's good to know that these options exist, but you won't use them this time.

6. Run the tool.

   Because Append added the Pacific_NW features to the Southwest layer, all states now draw using the Southwest colors.

7. Turn off the Pacific_NW layer to confirm that the Southwest layer has the Pacific_NW states appended to it.

Pairwise Clip tool

Clip extracts the input features that are overlaid by the clip features. This tool is best thought of as a cookie cutter. Clip is particularly useful for creating a new subset feature class (also known as an area of interest) that contains geographic features from another, larger feature class. A clip polygon can be used on polygon, line, or point layers (as illustrated in each of the following images, respectively). Why would you use it? Most commonly, it's to reduce a feature to only your study area, whether that's a city, a postal code, or any other shape.

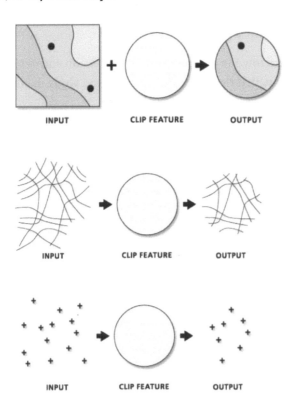

Use the Pairwise Clip tool to create a feature class

1. In the Contents pane, turn on the Pacific_NW layer and turn off the Southwest layer.

2. Click Add Data to add the US_counties layer from the Chapter14 folder.

3. In the Contents pane, drag the Pacific_NW layer above the US_counties layer.

4. On the Analysis tab, in the Tools gallery, locate the Pairwise Clip in the default group, and open the tool in the Geoprocessing pane.

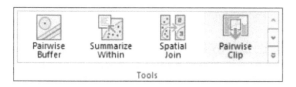

5. In the Pairwise Clip tool pane, for Input Features, add the US_counties layer.

6. For Clip Features, add the Pacific_NW layer.

7. For Output Feature Class, browse to the Chapter14 folder, and name the shapefile **US_counties_PairwiseClip**.

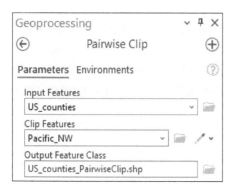

A new layer will appear of only those counties within the boundaries of the Pacific_NW states (the cookie cutter).

8. Run the tool.

   It's a bit difficult to see what happened in your map view.

9. Turn off the Pacific_NW and US_counties layers to better see the US_counties_PairwiseClip layer.

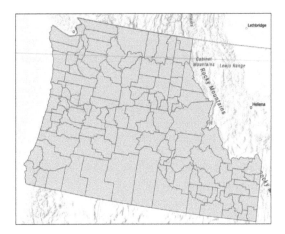

Now you have all the counties for those two states, with all their original attributes, in their own layer.

Hint: Open the attribute table to confirm whether that's true.

How to search for a tool

Although ArcGIS Pro makes a collection of popular tools—Favorite tools and Recent tools—easily accessible, you'll inevitably need a tool that isn't collected in those groups. And digging through all those toolboxes can be a real pain. Luckily, there's also a great search function called Find Tools at the top of the Geoprocessing pane. You can get to it by clicking Tools on the Analysis tab in the Geoprocessing group.

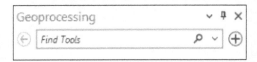

As you type, ArcGIS Pro dynamically starts guessing which tool you want and populating the Geoprocessing pane with choices. Each tool has a brief description of what the tool does to help you decide. When you see the tool you want in the list, you can click it to launch the tool.

After you run a tool, it will appear in your Recent list under Favorites in the Geoprocessing pane. From there, you can even add the tool to your favorites (right-click the tool and click Add to My Favorites).

Optional exercise: start typing "intersect" in the Find Tools text box to see how the search works. No need to run the tool. Of course, if you're feeling adventurous, go for it and see what happens. Come on, we dare you!

Pairwise Dissolve tool

The Pairwise Dissolve tool aggregates features based on specified attributes. The aggregated output features can be summarized or described using a variety of statistics. Users can choose to add a Sum field, for example, to compute the population totals for all smaller input areas in each of the respective larger output areas, as shown in these images.

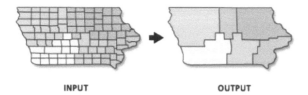

INPUT   OUTPUT

Use the Pairwise Dissolve tool to create an aggregated feature

1. Turn off all layers in the Contents pane.
2. Click Add Data to add the States layer from the Chapter14 folder.
3. Open the attribute table for the States layer to confirm that there's a field called Region.

   Region designates which region of the country each state is in.
4. Close the attribute table.
5. Open the Symbology pane for States, and for Primary Symbology, click Unique Values.
6. For Field 1, choose Region.
7. On the Classes tab, click Add All Values to display the regions symbolized with random colors.

All the states are now symbolized by their respective regions. What if you wanted a layer in which all the states in a given region were combined, giving you four features, one for each region? It's Pairwise Dissolve to the rescue!

And how handy that you just learned about searching for a tool (in "How to search for a tool"), because the Pairwise Dissolve tool is not in the Tools gallery.

8. On the Analysis tab, in the Geoprocessing group, click Tools to open the Geoprocessing pane.

9. In the Geoprocessing pane, in the Find Tools text box at the top, type **dissolve**.

The tool you want is at the top of the list: Dissolve.

10. Open the Dissolve tool.

ArcGIS Pro alerts you that a Pairwise option for the Dissolve tool is available. Let's use it.

11. At the top of the tool pane, click Pairwise Dissolve to open the Pairwise Dissolve tool.

12. In the Pairwise Dissolve tool pane, for Input Features, add the States layer.

13. For Output Feature Class, browse to your Chapter14 folder, and name the shapefile **US_regions**.

14. For Dissolve Fields, click the Region field from the list.

15. Run the tool.

    When the tool finishes, not only is your new output feature added to the map, but it's also symbolized in the same colors that were specified for the regions in the original input States layer. How nice!

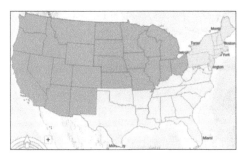

16. Save and close your project.

    We'll admit this was a long chapter! But you just learned how to use five different geoprocessing tools, as well as learning the Measure tool and how to drag layers into the tools. That's a lot! You've earned a break. Go relax and enjoy your favorite snack.

USER STORY

Mapping helps protect unsheltered people from wildfires

Knowing where homeless populations are sheltering is vital during wildfire evacuations, and students from Anderson W. Clark Magnet High School in Los Angeles County have done something to help. The numbers of both wildfires and people who are homeless have grown in California. The forests, fields, and hills in urban areas are full of dry brush because of the drought. These students set out to prove that infrared images analyzed with GIS technology could identify encampments of people who are homeless near areas at high risk of wildfire.

The students searched for a test site, performing a site suitability analysis using factors such as land type, locations without drone flight restrictions, and fire risk based on Los Angeles County data. In a GIS, they layered the data to identify a high-risk area within a 15-mile radius of their school. The resulting intersected layer of encampments of people who are homeless near areas at high risk of wildfire are the green areas on the map.

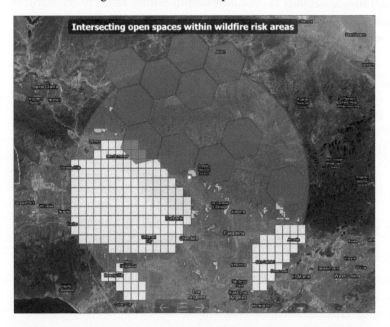

Map by Los Angeles County high school students pinpoints where people who are homeless may be camping in areas at risk of wildfire.

Credit and data source: Anderson W. Clark Magnet High School, County of Los Angeles.

CHAPTER 15

Creating geodatabases

By now, you're familiar with shapefiles, but there's another, more advanced type of GIS data storage. In this chapter, you'll learn how to create a geodatabase, which is a container for spatial and attribute data that can store different types of GIS data in one place. This all-encompassing structure can contain all the point, line, polygon, raster, and tabular data you'll need. Having all your GIS data stored uniformly in a central location allows for easy access and management. Is that exciting stuff? Will it make your work easier when you're using a variety of data? Of course, and then you'll thank us for the time you spent learning about it.

Why might knowing about a geodatabase be important? The geodatabase is the industry standard format for managing GIS data. As a working GIS professional, you'll probably store your GIS data in a file or enterprise geodatabase format. You'll focus on file geodatabases for now.

Explore the Catalog pane

As with so many tasks in ArcGIS Pro, you have multiple ways to make a new file geodatabase. You can use the Catalog pane or a geoprocessing tool. It's a good idea for you to become familiar with the Catalog pane, so we'll go that route.

1. In ArcGIS Pro, start a new map project. Name it **Ch15_geodatabase**, and save it to your Chapter15 folder.

2. On the View tab, in the Windows group, click Catalog Pane to open it to the right of the map view.

   The Catalog pane appears, showing a list of databases, maps, toolboxes, styles, folders, and even locators for geocoding, all of which can be managed in the pane. You're just starting out with a new project, so the pane won't contain much yet. That's okay.

Create a file geodatabase

1. In the Catalog pane, right-click Databases, and click New File Geodatabase.

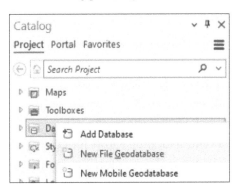

2. Save the new geodatabase to your Ch15_geodatabase folder, name it **MyGDB**, and click Save.

Chapter 15: Creating geodatabases

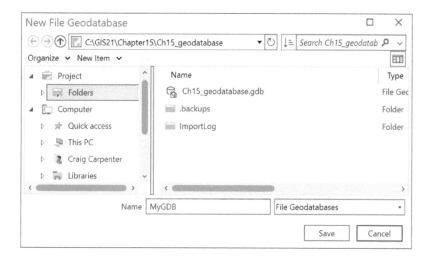

The little progress bar spins around for a bit, but it looks as if nothing happened. Not so fast!

3. In the Catalog pane, expand Databases.

   MyGDB.gdb is now visible. Cool!

Ch15_geodatabase is also the name of your project. That geodatabase is automatically made when you create a project, but that isn't nearly as fun as making a geodatabase on your own. If you expand MyGDB.gdb, you'll see there's nothing in it. Until you add some data to the geodatabase, it's just an empty container. Can you guess what happens next?

Import shapefiles into a geodatabase

1. In the Catalog pane, right-click MyGDB.gdb, and click Import > Feature Class(es).

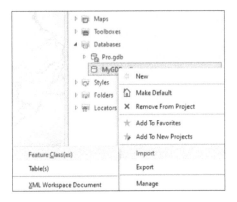

The Feature Class to Geodatabase tool pane opens. Because you opened it by right-clicking your new file geodatabase, ArcGIS Pro autocompletes the parameter for Output Geodatabase.

2. For Input Features, browse to your Chapter14 folder, and click States.shp to add that shapefile to the list.

A new input text box for Input Features appears.

3. In the new text box, browse to your Chapter4 folder, and click CountiesAge.shp.

You'll import only those two shapefiles.

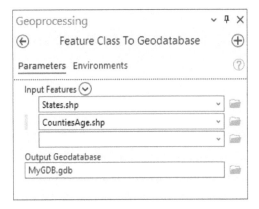

4. Click Run to run the tool.

Remember the progress bar at the bottom of the Geoprocessing pane and the pop-up summary of the tool you're running? When the tool has finished, you can click View Details under Run in the pane to see what happened. You were successful again! These details are even more useful if anything goes wrong.

5. To verify that the two shapefiles were added to your new geodatabase, click the Catalog tab to return to the Catalog pane.

6. Right-click MyGDB.gdb, and click Refresh.

   Refresh updates the contents to show recently added items. That's exactly what you want!

   The shapefiles have indeed been copied into your geodatabase. Yay!

   Refresh your geodatabase view if you're looking for something but not seeing it.

   When you imported these shapefiles, you converted them into a new format called a feature class, which is an improved way of storing the spatial data from the shapefiles.

Import a table into a geodatabase

Many other data types can be imported into a geodatabase. Let's do a table next!

1. In the Catalog pane, right-click MyGDB.gdb, and click Import > Table(s).

   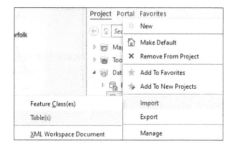

The Table to Geodatabase tool pane appears, with Output Geodatabase autocompleted.

2. For Input Table, browse to your Chapter03 folder, and click the Seniors.xlsx > Seniors$ sheet.

   An additional text box for Input Table appears, if you wanted to import multiple tables, but you'll add only the one table. Good to know that you can add multiple tables at a time.

3. Click Run to run the tool.

4. Click the Catalog tab to return to the Catalog pane.

   Weird. Where is your table in MyGDB.gdb?

5. Right-click MyGDB.gdb, and click Refresh.

   Sure enough, there's your table. Whew!

What if I don't see my geodatabase in the Catalog pane?

ArcGIS Pro doesn't necessarily save all your database connections from project to project. If you don't see your geodatabase listed in the Catalog pane in your next project, no need to panic! You can right-click Databases in the Catalog pane, click Add Database, and browse to the GDB file you want to see. Easy!

Load an aerial photo into a geodatabase

Aerial imagery, which you'll use in a later chapter, is handled a little differently in a geodatabase.

1. Open the Geoprocessing pane.

2. Click the back arrow if needed. In the Find Tools search box, type **raster to geodatabase**.

   You don't need to type the whole phrase for the search function to locate the tool. The tool you need is right at the top.

Chapter 15: Creating geodatabases

3. Click Raster to Geodatabase to open the tool.

4. For Input Rasters, browse to the Chapter15 folder, click the USC.tif file, and click OK.

5. For Output Geodatabase, browse to MyGDB.gdb in the Ch15_geodatabase folder.

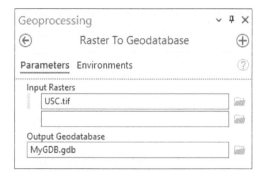

6. Run the tool.

7. When the tool has finished running, click the Catalog tab to return to the Catalog pane.

    Remember to refresh if you need to. Yep, there's your raster in the geodatabase.

    Good work! Now you know how to create a geodatabase and add different data to it. This skill will come in handy in later chapters.

8. Save and close your project.

Designing a proper geodatabase

What we've done here was simply add data into a geodatabase, which is fine for learning the basics. But you should know that a geodatabase is commonly made up of feature datasets (larger data categories), which hold the actual feature data, called feature classes, as well as numerous other elements (refer to the image). Part of what makes geodatabases the industry standard format is that they can store varied data types within one file. That capability may not seem important, but without it, after a few projects, you would have quite a mess of data to dig through to find what you need.

For large systems, designing a proper geodatabase structure can be a task. In fact, entire books have been written about designing geodatabases, including *Focus on Geodatabases in ArcGIS Pro* by David W. Allen (Esri Press, 2019). Lucky for you that we didn't make you do anything that complicated!

USER STORY

Exploring the National Hydrography Dataset

The US federal government is an important source of geospatial data. Here, we highlight the National Hydrography Dataset (NHD) compiled by the US Geological Survey (USGS). The NHD represents the water drainage network of the United States with features such as rivers, streams, canals, lakes, ponds, coastlines, dams, and stream gauges. The NHD is the most up-to-date and comprehensive hydrography dataset for the nation.

The National Map Download viewer provides an online mapping tool for downloading the NHD file geodatabase. Check it out at apps.nationalmap.gov/downloader.

To learn more about the National Hydrography Dataset, visit links.esri.com/NHD.

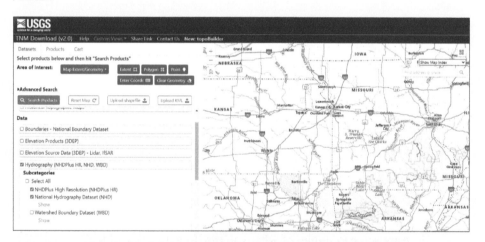

The NHD is a download of the national geodatabase of the country's hydrography dataset.

Credit: USGS, US Department of the Interior, National Hydrography Dataset (NHD) Downloadable Data Collection, National Geospatial Data Asset (NGDA), National Geospatial Technical Operations Center (NGTOC).

CHAPTER 16

Joining features

In an earlier chapter, you learned how to join layers of data by matching a shared tabular attribute. In this chapter, you'll learn how to join the attributes of data layers by the location of their features, referred to as a spatial join. A spatial join matches rows from the join features to the target features based on their locations. Yes, it may sound like a bunch of buzzwords, but it will make more sense when you do it.

Here are some examples of what you can do with a spatial join:

- Add postal codes or census tracts to point features.
- Add city council or congressional districts to projects.
- Count the number of projects in an area.

All very practical stuff, right?

In this case, you'll add an .lyrx file instead of a .shp file because it has some complex symbology that's useful here. An .lyrx file saves the symbology, labeling, and other specifications that were made previously.

Add data

1. In ArcGIS Pro, start a new map project. Name it **Ch16_spatialjoin**, and save it to your Chapter16 folder.

2. Click Add Data, and first add the CaliforniaCounties.shp from the Chapter01 folder, and then add the Storm_Events.lyrx from the Chapter16 folder.

3. In the Contents pane, if the Storm_Events layer is not listed above the Counties layer, drag the Storm_Events layer to the top so you can see its features.

Select storm events for only your state

The Storm_Events layer includes storm data for the entire United States. That's more than 60,000 points! Let's narrow that down so the spatial join will run faster.

1. On the Map tab, in the Selection group, click Select by Location.

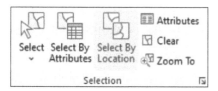

2. In the Select by Location tool pane, for Input Features, confirm that the Storm_Events layer is selected.

3. For Relationship, keep Intersect.

   But take a look at some of the other options available. There are many handy options for another time.

4. For Selecting Features, click the CaliforniaCounties layer to select all the storms for your state only.

5. Click OK.

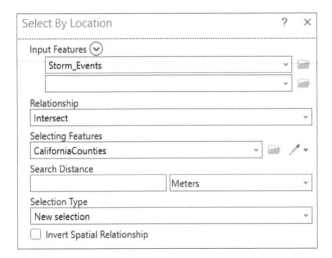

Export the selected points

The storm points are selected in the map. You'll make that selection its own layer.

1. In the Contents pane, right-click the Storm_Events layer, and click Data > Export Features.

   Now that you know about geodatabases and feature classes, instead of saving the output as a shapefile in a folder, you'll put your new skills to use and save the output as a feature class in the project's geodatabase.

2. In the Export Features tool pane, click inside the Output Feature Class text box with the default name. Widen the pane so you can see the entire file path.

   By default, your output feature class is saved to the Ch16_spatialjoin.gdb. That seems like a perfect place!

3. Carefully rename the file name at the end of the file path to **Storm_Events_CA**.

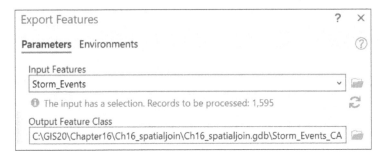

Chapter 16: Joining features

4. Click OK to save your state's storm event points as a feature class in your project geodatabase.

   ArcGIS Pro adds the new layer to the Contents pane.

5. In the Contents pane, turn off the Storm_Events layer for the entire country so the new layer is visible.

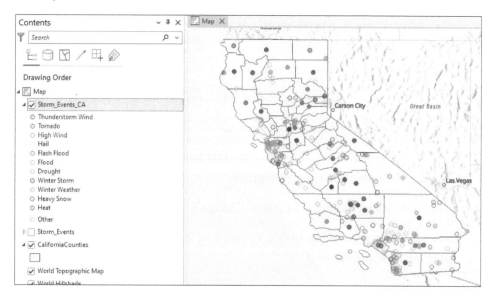

Perform a spatial join

Spatial joins can go either way between the two layers. We can add information to our county polygons from the storm event points or add information to our storm event points from the county polygons. We're going to perform a spatial join in both directions to demonstrate the difference.

Add county data to the storm event points

1. On the Analysis tab, click Tools.

   How about that? The Spatial Join tool is even listed on the Favorites tab. (If not, just search for it in Find Tools.)

2. Click Spatial Join to open the tool.

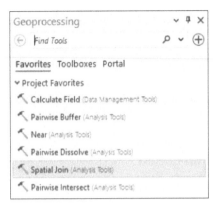

3. In the Spatial Join tool pane, for Target Features, add Storm_Events_CA.

4. For Join Features, add CaliforniaCounties.

5. For Output Feature Class, click inside the text box, and rename the file at the end of the file path **Storm_Counties_SpatialJoin**.

6. The default Match Option is Intersect. Change it to Within.

   Each storm event point lies within a county, so the corresponding county's attributes will be appended to it.

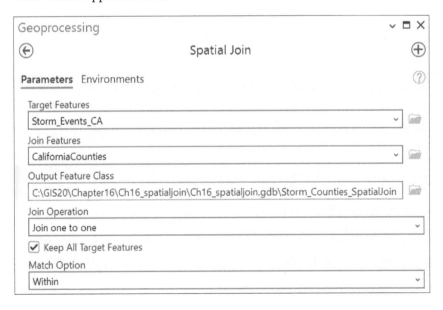

7. Run the tool.

When the Spatial Join tool finishes, the new Storm_Counties_SpatialJoin layer is added to both the Contents pane and the map with the same symbology as that found in the layer of storms for your state.

Examine the attribute table

The new point layer should have attributes from the counties polygons. Don't believe us? Let's check it out.

1. Open the attribute table of the new Storm_Counties_SpatialJoin layer.

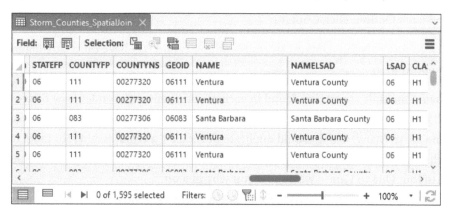

Sure enough! The attribute table of the new layer has the fields and attributes from the County polygon layer, including the County name for each storm event.

2. Close the table.

Now let's perform a spatial join in the other direction, adding the storm event attributes to the county polygons. This approach can be useful for knowing how many storms happen in each county.

Perform a second spatial join

Add storm event data to the county polygons

1. If the Spatial Join tool isn't still open, reopen it.
2. In the Spatial Join tool pane, for Target Features, add CaliforniaCounties.
3. For Join Features, add Storm_Events_CA.
4. For Output Feature Class, name it **Counties_Storm_SpatialJoin**.

5. For Match Option, click Within.

   A white-on-red *X* appears, indicating a problem.

   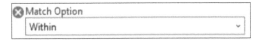

   If you click the *X*, a full explanation appears describing why this relationship is invalid. Essentially, a polygon cannot be within a point, so Within is an invalid choice. Makes sense. Let's choose something else, instead.

6. For Match Option, click Contains.

   Each county polygon contains storm event points and will have those attributes appended to it.

   Because multiple storms are found in each county, you may wonder how this join will work. For Join Operation, you have a choice of two options. You can choose Join One to Many, which would duplicate the county attributes and join them to each storm that occurred in a given county. But for now you'll keep Join Operation set to Join One to One—we'll show you why in a bit.

7. Run the tool.

When the Spatial Join tool finishes, the new Counties_Storm_SpatialJoin layer is added to both the Contents pane and the map with the same symbology as that used in the CaliforniaCounties layer.

Examine the attribute table

The new county polygon layer should have attributes from the storm event points. We bet you'll believe us this time!

1. Open the Counties_Storm_SpatialJoin attribute table, and find the Join_Count field.

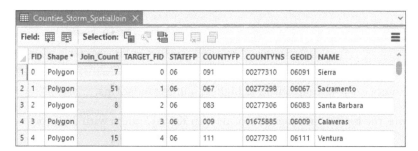

Now you understand why you left the Join Operation parameter set to Join One to One. The spatial join added the attributes for only the first storm it encountered for each county, but it added the Join_Count field, which lists the total count of all the storms in each county. From here, you could easily make a choropleth quantitative map showing the counties in a color range for the number of storms. In fact, let's do that!

2. Close the attribute table.

Map the storm counts by county

1. In the Contents pane, turn off and collapse all the layers except Counties_Storm_SpatialJoin.

2. Right-click the Counties_Storm_SpatialJoin layer, and click Symbology.

3. In the Symbology pane, for Primary Symbology, click Graduated Colors.

ArcGIS Pro chooses a random color scheme. You can now easily tell which counties have the most storms. For California, there's a deep-red county near the center. Wonder which county that is? Let's find out.

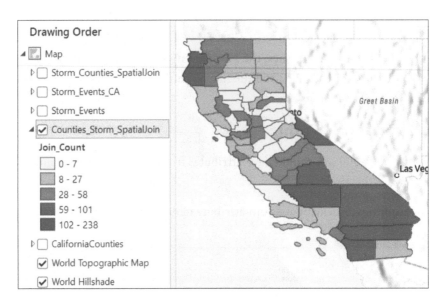

4. Click the county on your map with the highest number of storms.

   In California, the highest number of storms is in Kern County, with 238 storms.

5. Save and close your project.

   The Spatial Join tool is pretty cool. It's powerful. You'll be glad you learned how to use it!

USER STORY
Finding vulnerable populations

Spatial joins can be useful in compiling a score for a location based on multiple variables. Check out how the Centers for Disease Control and Prevention (CDC) developed a Social Vulnerability Index.

Public officials responding to the coronavirus disease 2019 (COVID-19) urgently tried to find and protect the most vulnerable among us. Vulnerability is determined by factors such as age, living situation, income, and preexisting health problems.

The CDC's Social Vulnerability Index uses a composite of 15 social factors, including poverty, lack of vehicle access, and crowded housing conditions. The index can flag communities that may experience a demand for health care that outpaces the local medical infrastructure.

To read more about the CDC's Social Vulnerability Index, see "Reveal: Finding the Most Vulnerable among Us" by Makenna Sones in *Esri Blog*, April 2, 2020, at links.esri.com/CDCVulnerability.

The Social Vulnerability Index map shows disposable income, poverty, and savings in New York households.

Credit: Centers for Disease Control and Prevention (CDC). Data source: CDC, US Census Bureau, NYC Open Data, New Jersey Office of GIS, Esri, HERE, Garmin, SafeGraph, METI/NASA, USGS, EPA, NPS, USDA.

CHAPTER 17

Working with imagery

You've already learned about GIS vector data types (points, lines, and polygons). In this chapter, you'll learn about raster data. A raster is just another way of representing data by using an image—data that's made up of a collection of pixels. Like an image on a camera or TV screen, the smaller the pixel size, the greater the resolution of the image. Satellite and modern aerial imagery tend to have relatively accurate location information but may need some adjustments to line up with all your GIS data. Historical imagery, which can be extremely useful, is another story and takes a few steps. Luckily, the software does the vast majority of the work.

Rasters can be viewed, queried, and analyzed in ArcGIS Pro. In this chapter, we'll show you how to bring raster data into GIS. To put it more simply, we'll show you how to transform a photo or drawing into a layer in a map.

Aerial photos also can contain a lot of data. As the saying goes, a picture is worth a thousand words. Being able to place an aerial photo or drawing in your map gives you a snapshot in time that you can combine with other data. Warning: Once you know how to do this and find data sources, you can spend countless hours with historical imagery. You know we have!

Add image data

1. In File Explorer, browse to C:\GIS20\Chapter17, and double-click Ch17_imagery.aprx to open the project.

   This project has already been created and prepared for you. It has a Streets layer and a bookmarked location of Washington, Illinois.

2. Click Add Data, browse to the Chapter17 folder, and add tornado.tif.

   This satellite image reveals the damage from an EF4 tornado (166–200 mph) that struck Washington, Illinois, on November 17, 2013.

   A pop-up message asks whether you want to calculate statistics for the image. You want your image to display as quickly and crisply as possible.

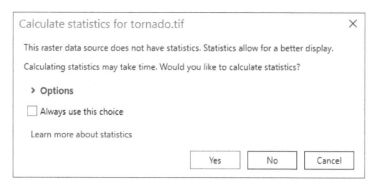

3. In the Calculate Statistics message, click Yes. If you're feeling ambitious, you can check the Always Use This Choice box first.

   The image layer is added to the Contents pane, but it won't draw on the map. Weird, right? Well, not really. The image doesn't have any location information yet, so ArcGIS Pro doesn't know where to correctly place it in the map. That's where you come in!

   You need to align, or georeference, the image. This georeferencing process will define the image's location on the map.

Georeference the image

1. In the Contents pane, make sure that tornado.tif is selected (it should be highlighted in blue), or click it to do so.

2. On the Imagery tab, click Georeference to open the Georeference tab on the ribbon.

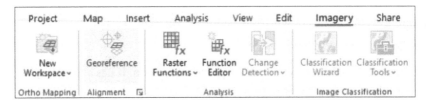

The tools on the Georeference tab are arranged into groups to help you use the correct tools in the different phases of your georeferencing session: Prepare, Adjust, Review, Save, and Close.

As you may have guessed, you're going to start in the Prepare group. Before you continue, though, you may have seen a pop-up in the upper right of the map frame.

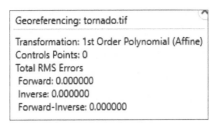

There's no information, but that's okay, because you certainly haven't done anything wrong. ArcGIS Pro is giving you the current status of the image, which is that no georeferencing has occurred.

3. In the Prepare group, click Fit to Display to place the image roughly within the current extent.

The image won't line up correctly with the streets yet. That's what you're going to do in the following steps. The satellite image clearly shows the path of the tornado as a white line.

> Your image appears properly oriented. But in the future, you can use Move, Scale, and Rotate in the Prepare group to orient the raster as desired.

4. Choose a color to make the streets easy to see. The examples use Ginger Pink, but feel free to use any color that works for you. Increase the line width, if you want.

Hint: To increase the line width, right-click the Streets layer in the Contents pane, and click Symbology. Click the line symbol, click the Properties tab, and change the line width.

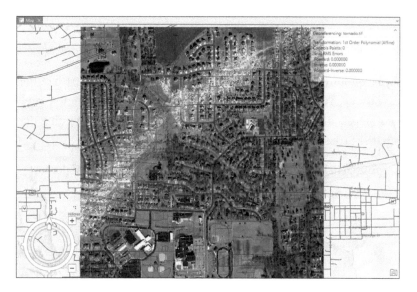

Add control points

Georeferencing usually requires at least three control points. Control points always occur in pairs: one point on a feature on the image, and then a second point on the vector layer to indicate that those two control points belong in the same place. The image is adjusted after each control point pair is added, so you may notice some skewed results when you add the first control points. Don't worry. Those results only indicate that you need to add more control points to get the image to match better. If at any point you feel that things have gone askew, you can click Reset in the Adjust group to clear all control points, and then click Fit to Display again to start over from scratch.

When choosing control points, try to look for well-defined objects in your images, such as road intersections or land features. They should be features on the ground, not elevated features. This way, you can be sure that you're referencing the same ground location in both the image and the target layers.

Hint: You can turn the image off and on to get perspective with the other layers.

Let's find a good location for our first control point.

> **Hint:** Use Navigator, your mouse, or the Map tab to locate a good spot for setting control points, and then click the Georeferencing tab when you're ready to add the points.

1. Zoom to the center-left of the map, near the building with the dark roof.

   To the left of the building with the dark roof, the intersection in the path of the tornado is easy to see. You'll add a control point to that intersection in the satellite image and georeference (move) the image to the correct location by finding that intersection in the Streets layer.

2. In the Adjust group, click Add Control Points to start creating control points.

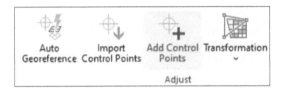

3. Review the image to see where to add the first control point pair. Click the location shown with a blue cross to add a control point to the image (the source layer), and then click the location shown with a red cross to add a control point to the Streets layer (the target layer).

   > *When you clicked Fit to Display, the placement of the satellite image may have been somewhat different because of your screen size. When placing the control points, refer to the image to choose the location of the control point for the source layer (blue cross), and refer to the streets to choose the location of the control point for the target layer (red cross).*

The image moves based on the control point pair you added. For the second control point pair, you'll use a street intersection to the north.

4. Review the figure to see where to add the second control point pair. Click the location shown with a blue cross to add a control point to the image (the source layer), and then click the location shown with a red cross to add a control point to the Streets layer (the target layer).

Again, the image moved based on the control point pair. You're getting it closer to the right place, but your zoom level is set rather oddly now, so you'll zoom out and spread your control points more evenly through the entire image to help you get the best fit for the image. Your two control point pairs are shown as red crosses in white circles on the map.

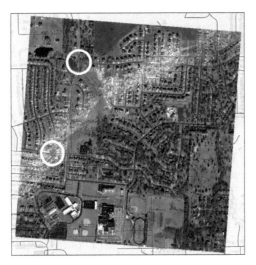

Hint: If you click a wrong point, in the Review group, click Control Point Table. Delete any unwanted control points using the table. Alternatively, edit the inaccurate points by selecting them and moving the vertices.

For the third control point pair, let's use a spot somewhere on the right side of the image.

5. Zoom out as needed.

6. Review the image to see where to add the third control point pair. Click the location shown with a blue cross to add a control point to the image (the source layer), and then click the location shown with a red cross to add a control point to the Streets layer (the target layer).

7. Zoom out a bit to see that the image fits well with the Streets layer.

   If you're a perfectionist (you know who you are!), feel free to add control points until you're satisfied. For the rest of you, proceed to the next step.

8. On the Georeferencing tab, in the Save group, click Save to save all your hard work.

   The Save button turns gray to indicate that your georeferencing has been saved.

9. On the far right of the Georeferencing tab, click Close Georeference.

10. Save and close your project.

    When you look at the tornado.tif file in File Explorer, you notice that a couple of friends have moved in. The .aux.xml file was created at the beginning when you calculated statistics. The .tfwx file was created when you clicked Save. This file tells ArcGIS Pro where to draw the image on the map.

    ```
    tornado.tfwx
    tornado.tif
    tornado.tif.aux.xml
    ```

    Pretty cool stuff, right? Imagery can be important for visualization, as well as for raster analysis techniques in GIS that are too advanced for this book. For example, raster images can be used for land-cover analyses, for change detection over time, or for identifying permeable surfaces to calculate storm water runoff. If things like that interest you, you may want to stick with GIS and continue improving your skills. Either way, you now have experience with georeferencing an image. Nice work!

Scanned documents are rasters, too!

Raster data can be obtained from many sources. Scanned maps and historical photos can be great information to combine with your vector layers or with other imagery. Scanned documents don't contain spatial reference information, so they must first be georeferenced to combine with other data.

One handy use is to scan and georeference as-built engineering plans so they can be used to digitize underground features, such as water lines, sewer lines, or storm drains. Almost every agency has drawings that would be more useful when brought into GIS. Some interesting examples of old maps have been scanned and georeferenced to show change over time. A great example: maps of Boston showing that much of what is now land was once part of Boston Harbor. Check out the David Rumsey Map Collection online at davidrumsey.com for this and other uses of historical maps.

Chapter 17: Working with imagery

USER STORY
Using the Swipe tool for disaster imagery

One powerful use of aerial imagery is to show the effects of natural disasters, such as hurricanes, tornadoes, earthquakes, wildfires, or flooding. A handy way of showing before-and-after imagery is by using the Swipe tool in ArcGIS Pro. In this example, we show imagery before and after Hurricane Irma's devastation in Florida in 2017.

This app was developed by the Esri Disaster Response team and is designed to be used interactively. Go to links.esri.com/HurricaneSwipe to try it.

The Hurricane Irma Post-Event Imagery Swipe Map allows users to visualize a section of Florida before (*left*) and after (*right*) hurricane flooding in 2017.

Credit: Esri Disaster Response Team. Data sources: NOAA, Esri Community Maps Contributors, FDEP, OpenStreetMap, Microsoft, HERE, Garmin, SafeGraph, GeoTechnologies Inc., METI/NASA, USGS, EPA, NPS, US Census Bureau, USDA, NASA, NGA, FEMA.

CHAPTER 18

Using 3D data

ArcGIS Pro was designed to easily handle 3D projects. In fact, 3D is an integral part of ArcGIS Pro. So we wanted to give you at least a taste of its 3D environment.

3D maps, called scenes, can be rendered as either global or local. Global scenes calculate for the curvature of the earth, and local scenes use local coordinates. Most of your work is probably best suited to local scenes, which allow users to focus on smaller project areas, such as viewing a community.

Any 2D map can be converted into a 3D map, so that you can visualize, explore, and analyze your data on a different level. A local 3D perspective brings real value to topics such as building heights, viewsheds, shadowing, sea level rise, or any belowground features, such as pipes or geologic features.

In addition to real-world features, features based on statistical attributes can be extruded into 3D features for a truly futuristic display. Perhaps the future is now! These images illustrate counts of public transportation ridership symbolized in 3D and an advanced 3D technique using voxels (cube representations of data) to show patterns of social distancing during the COVID-19 pandemic.

Chapter 18: Using 3D data

 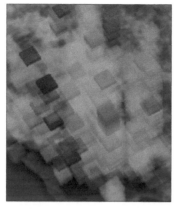

Left: Example of 3D extrusion illustrating a new rail line in San Francisco, California, with transportation ridership. The higher the column along each section of the line, the greater the ridership. *Right*: Example of voxel analysis rendered in 3D. Behavioral patterns for social distancing are illustrated as a multidimensional voxel layer.

Source: *Left*: ArcGIS tutorial, "Visualize the Expansion of Public Transportation," at https://learn.arcgis.com/en/projects/visualize-the-expansion-of-public-transportation. *Right*: ArcGIS tutorial, "Visualize Social Distancing across California," at https://learn.arcgis.com/en/projects/visualize-social-distancing-across-california.

Pretty cool, right? Now, we're not going to get all fancy here, but if you enjoy this taste of 3D GIS, just know that there's plenty more 3D out there for you. Let's get started!

> Using 3D data can sometimes be a drain on your computer, so remember to save often.

Convert a 2D project into a 3D project

1. In File Explorer, browse to C:\GIS20\Chapter18\Ch18_3D_data, and double-click Ch18_3D_data.aprx to open the project.

   The map shows 2D building footprints in the Rutland, Vermont, area. How would they look in 3D? Let's find out!

2. On the View tab, in the View group, click Convert > To Local Scene.

This conversion takes a moment, and the result doesn't look much different. ArcGIS Pro kept your 2D map on its own tab and added a tab called Map_3D with a local scene icon. The 3D scene has copies of the data layers from the 2D map, and a new layer was added in the Contents pane: WorldElevation3D/Terrain3D. This layer is the 3D surface that acts as the ground layer for your other layers.

Let's move around the new 3D scene to view it.

3. On the Map tab, in the Navigate group, click Explore to activate it.

4. To explore the scene, click and hold the scroll wheel while moving the mouse around. Use the on-screen Navigator tool to try camera navigation techniques such as rotating around a target point.

Improve the look of the 3D scene

That topographic basemap isn't suitable for highlighting the 3D effect, so you'll use another basemap.

1. On the Map tab, in the Layer group, click Basemap.

2. From the basemap choices, click Light Gray Canvas.

   Now you'll add shading to the elevation surface layer. You can play with the illumination of the scene to try different effects.

3. In the Contents pane, under Elevation Surfaces, click Ground to open the Elevation Surface Layer tab at the top.

   > **Hint:** If the tab is displayed but didn't open, click it to open it.

4. On the Elevation Surface Layer tab, in the Surface group, check the box for Shade Relative to Light Position.

   This option changes the scene, so be patient while you watch the progress icon on the bottom right.

5. In the Contents pane, right-click Map_3D, and click Properties.

6. In the Map Properties dialog box, click Illumination on the left.

7. Under Illumination Defined By, change the Altitude value to **65**.

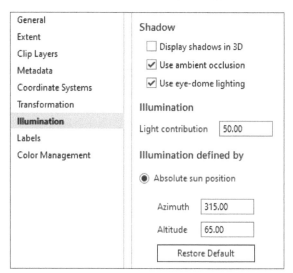

Let's change one more thing before we click OK.

8. In the Map Properties dialog box, click the General tab.

9. Click the Background Color down arrow. From the color scheme, pick a color to represent the sky.

10. Click OK.

Fewer shadows appear now that the sun is set at a higher altitude.

11. View the updated scene from different perspectives.

Change building symbology

1. In the Contents pane, right-click the BuildingFootprints layer, and click Symbology.

2. In the Symbology pane, on the Primary Symbology tab, click the color square for Symbol.

3. If necessary, click the Gallery tab.

   Hey, there's a color option called Building Footprint. ArcGIS Pro thinks of everything!

4. Under ArcGIS 2D, click Building Footprint to apply that style.

   Applying the Building Footprint symbology may take a moment. When the scene finishes drawing, the new symbology appears, but the buildings still look flat. You're going to use height values in the attribute table to extrude the building footprints into 3D features.

Chapter 18: Using 3D data

Extrude the BuildingFootprints layer

First, you'll find the height field in the attribute table for BuildingFootprints.

1. Open the attribute table for the BuildingFootprints layer.

   The BuildingHe field (short for BuildingHeight) stores the height of each building in feet.

2. Close the table.

3. In the Contents pane, make sure that the BuildingFootprints layer is selected.

4. On the ribbon, click Feature Layer to open that tab.

5. On the Feature Layer tab, in the Extrusion group, click Type > Max Height.

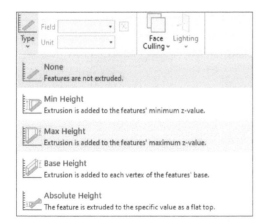

The process takes a moment for the BuildingFootprints layer to be added to the 3D Layers section of the Contents pane. Keep an eye on the progress icon to know when to proceed.

6. Click Save to save your work so far.

7. In the Extrusion group, click the Field down arrow, and click BuildingHe.

8. For Unit, click US feet.

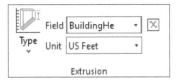

No need to click an Apply button. ArcGIS Pro immediately starts extruding those buildings. It may take a little time, so monitor the progress icon.

9. When the scene finishes drawing, view the updated scene from different perspectives.

Wow, that's cool! And you did it all by yourself! Well, with a little help from us and that field with the building heights. But it was really easy, right? You learned how to create a 3D scene, change the illumination settings, and create 3D features by extruding with a height field. You'll never look at a video game the same way again!

If you're hooked now and want to read about more advanced 3D data that can be used in GIS, see the lidar section and user story.

10. Save and close your project.

Lidar

Lidar is an abbreviation for light detection and ranging, a useful technology we can use in GIS. Lidar works much like radar, but instead of using radio waves, it uses laser pulses. The time for the pulse to reflect to the sensor gives a range that's converted into the position of the point, including height. Lidar data can create a 3D model of an area. Even better, it can create multiple layers! For example, lidar can give you a terrain model of the bare earth, buildings, and the tree canopy.

Lidar data is generally obtained by aircraft and drones. Gathering lidar data is surprisingly inexpensive compared with gathering aerial photography and processing that photography to create 3D data. In many cases, lidar and aerial photos are gathered at the same time and are a good complement to each other.

A few years ago, using lidar was something of a challenge, but the software and processing power of the hardware have caught up. The US Geological Survey (USGS) is a great source for free lidar data, although if you need current data for an area, you'll need to pay to have it gathered or use a drone with a lidar system and do it yourself. For a single project, that's a bit of an investment, but it can pay off if you plan to use it often.

USER STORY
Using lidar to map hurricane flooding

The USGS relied on lidar technology in select areas of Louisiana, Mississippi, and Alabama to map flooding in urban areas affected by Hurricane Isaac in 2012. The image shows a 3D lidar scan of the Interstate 510 bridge in New Orleans, Louisiana, following Hurricane Isaac, indicating flooded areas along the river. Images such as this are capable of separating the data into layers, allowing officials to analyze each part separately. The bare earth layers are particularly useful.

For more information, visit links.esri.com/LidarHurricane.

A 3D terrestrial lidar scan of the Interstate 510 bridge in New Orleans after Hurricane Isaac flooded the waterfront in August 2012.

Source: USGS.

CHAPTER 19

Adding a table and chart to a layout

The maps you've made so far are probably perfect. But sometimes a little addition can make a map even more perfect. What's that? Something can be more perfect? That's because you haven't tried adding a table or a chart to your map. In this chapter, you'll learn how to add a table and chart to your map layout. Tables and charts add another level of understanding to a map. Because people learn in different ways, a table or chart may not be helpful for you but may be just what someone else needs to better understand your map's message.

Your map may show a variable symbolized with ranges, but a table allows users to see the exact value of the variable for a feature of interest. This can help them better understand it in a way that a legend can't convey. Likewise, a chart, for example, allows users to see the number of counties on your map within various ranges and may help them better understand the distribution of the variable among the counties—adding a level of understanding about the underlying data. Is a particular county in the same range as other counties? In a range of fewer counties? Don't worry if this sounds complicated—we'll show you. Tables and charts aren't only useful, they're also easy to compile and manage.

Insert a table frame

Let's start by opening an existing project and adding a table.

1. In File Explorer, browse to C:\GIS20\Chapter19, and open the Ch19_charts project.
2. Click the Layout tab to activate it.

   This layout is similar to the one you created earlier in the book but uses the CountiesAge feature class stored in a geodatabase in the Ch19_charts project. If you want to use a map of a different state, use the skills you've learned to add a feature class to the Ch_19 charts geodatabase and replace the map frame of California with your state.

3. In the Contents pane, under Map Frame, click the CountiesAge layer to activate it.
4. On the ribbon, on the Insert tab, in the Map Surrounds group, click Table Frame.

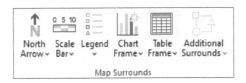

5. In the layout view, draw a box where you want the table to go, as indicated.

   A box draws first, with grab handles, and then the table from the Seniors layer populates the table frame.

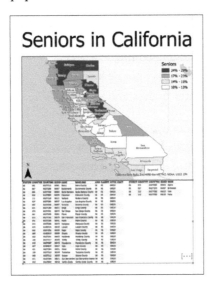

The table frame is a good start, but it's not showing the right fields in the table. Not so helpful. Good thing it's easy enough to change that.

Modify the fields in the table

1. In the Contents pane, under Layout, find the new Table Frame element.

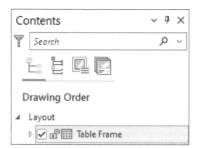

2. In the Contents pane, expand the Table Frame element.

   Only the first 10 fields from the layer are turned on and shown in the table. You'll adjust which fields are shown.

3. Uncheck all the fields except Name to turn them off.

   The fields disappear from the table as you turn them off. But you still don't have the option to add other fields. Remember how the field with the percentage of seniors was added at the end of the table when you created this layer using a join? Let's find out how to add that field to our table.

4. In the Contents pane, right-click Table Frame, and click Add Field > Seniors.

   The correct fields appear in the table. Next, you'll edit the Name field so it isn't notated in all-capital letters in the table and so that the numbers in the Seniors field have percentage signs after them.

5. In the Contents pane, right-click CountiesAge, and click Data Design > Fields.

6. Find the Name field in all capital letters in the Alias column, double-click inside the cell, and rename it **Name**.

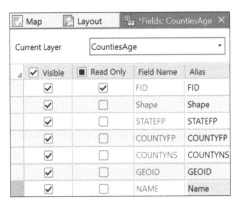

You know that the Seniors field is a percentage, so you'll change the formatting to add a percentage sign, making it easy for users to understand.

7. Find the Seniors field and the Number Format column, double-click the cell where they intersect (the text Numeric will be showing), and click the box in the cell.

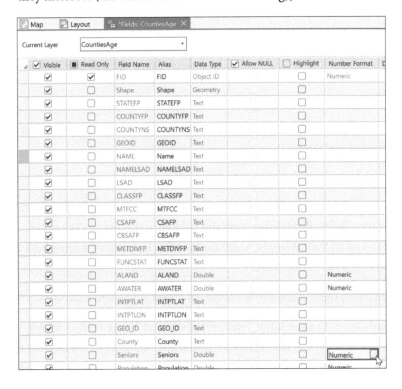

8. For Category, click the down arrow, change the selection to Percentage, and click OK.

9. On the ribbon, on the Fields tab, in the Changes group, click Save.

10. To exit the Fields view, click the X on the Fields tab.

   That looks better!

   > *If you used your own state with more counties than California, your table may show the More button (three horizontal red points in brackets), which means there is more data in your table that can't fit. You can customize all the font sizes and other table elements, but simply shifting the table and map sizes may do the trick.*

Sort the values of the Table Frame element

Let's make the table user-friendly by sorting the values.

1. In the Contents pane, right-click Table Frame, and click Properties.

2. In the Table Frame pane to the right, click the Arrangement tab.

3. Under Sorting, click the green plus sign, and click Seniors. Under Ascending, uncheck the Seniors box.

The table orders the percentage of seniors from highest to lowest. That looks better!

Name	Seniors	Name	Seniors	Name	Seniors	Name	Seniors
Alpine	29.1%	El Dorado	21.2%	Solano	15.7%	Santa Clara	13.5%
Sierra	28.2%	Shasta	20.6%	Ventura	15.6%	Stanislaus	13%
Mariposa	28.2%	San Luis Obispo	20.1%	Sutter	15.4%	Imperial	12.9%
Calaveras	28%	Tehama	19.8%	Santa Barbara	15.3%	San Joaquin	12.8%
Trinity	28%	Sonoma	19.6%	Mono	15.3%	San Benito	12.7%
Plumas	27.7%	Placer	19.6%	Orange	14.8%	Yolo	12.5%
Modoc	27.4%	Napa	19.2%	Colusa	14.6%	Yuba	12.4%
Nevada	27.4%	Butte	18.2%	Riverside	14.5%	Fresno	12.2%
Amador	27%	Humboldt	17.8%	Lassen	14.4%	San Bernardino	11.6%
Tuolumne	26.1%	Del Norte	17.8%	Sacramento	14.1%	Tulare	11.4%
Siskiyou	25.2%	Santa Cruz	16.4%	San Diego	14.1%	Merced	11.2%
Inyo	23.3%	Glenn	16.4%	Madera	14%	Kern	11%
Lake	22.8%	San Mateo	16.2%	Alameda	13.9%	Kings	10.3%
Marin	22.3%	San Francisco	15.8%	Los Angeles	13.6%		
Mendocino	22.1%	Contra Costa	15.8%	Monterey	13.6%		

4. Save the project.

That was cool. Let's take it to the next level and try the same thing with a chart this time. If you added a chart to the same layout to which you just added the table, the result would be cluttered. So you'll make a new layout based on this one.

Create a chart

1. On the ribbon, on the Layout tab, in the Page Setup group, click Duplicate Layout, and under ANSI Sizes, click Letter 8.5-by-11-inch.

2. In the Duplicate Layout dialog box, for Name, type **Chart**.

   A new layout tab named Chart is now active.

3. In the Chart layout, right-click the table, and click Delete.

   Don't worry—you're only making room for your chart.

4. In the Contents pane, under Map Frame, right-click CountiesAge, and click Create Chart > Histogram.

Chapter 19: Adding a table and chart to a layout

After you choose the chart type, a chart view appears below the map frame, but it's blank until you specify the chart's variables. The chart window is dockable, too, so you can move it to another place in the project, as you can with the Map, Layout, and Table views.

5. In the Chart Properties pane, on the Data tab, below Variable, for Number, click the down arrow, and click Seniors.

Your chart populates based on that variable. For now, you're going to keep it simple, but depending on your chart, you can choose other specifications in the future. In fact, ArcGIS Pro adds a line showing the mean of the data. We didn't even think of including that, but we like it, so we'll keep it.

Now that your chart looks good, you're ready to add it to the layout.

6. Close the Chart view.

Insert a chart

1. On the ribbon, on the Insert tab, in the Map Surrounds group, click the down arrow for Chart Frame, and click the Distribution of Seniors chart.

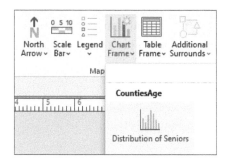

2. In the layout, use the guides to draw a box filling the bottom of the layout.

   It may take a few moments for the chart to appear in the frame. Also, the Chart Properties pane opens to the right, if you want to tweak the look of your chart. You'll make just one quick change.

3. In the Chart Properties pane, click the General tab. In the Chart Title text box, change the title to **Percentage of Seniors by County**, and press Enter.

   The chart title changes in the layout. The layout looks sharp with the map and the chart! Your layout should resemble this image.

   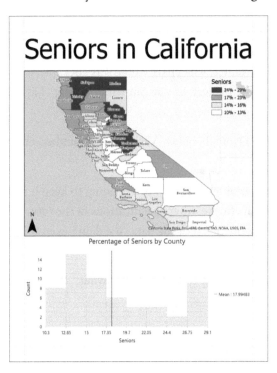

4. Save and close the project.

   Congratulations! You've learned how to add value to your map by inserting a table and a chart. Your map users will certainly appreciate having a map with tabular data, graphs, or charts together in one document. You can have your data and map it, too!

USER STORY
Using tables and charts to track pandemic data

The COVID-19 pandemic relied on maps, tables, and charts through a critical time. Researchers mapped COVID-19 case rates, hospitalization rates, death rates, and vaccination rates for many jurisdictional levels around the world. Officials even analyzed disease rates and outcomes across various racial and economic demographics.

The world leader in mapping COVID-19 is the Johns Hopkins University of Medicine in Baltimore, Maryland. The medical school analyzed near-real-time data from around the world to provide maps, such as the graduated symbols on the world map in the dashboard, and used that numeric (quantitative) data to add tabular elements and charts to the COVID-19 dashboard.

For additional information, see the Johns Hopkins University of Medicine Coronavirus Resource Center and COVID-19 Dashboard at coronavirus.jhu.edu/map.html.

The COVID-19 dashboard helped decision-makers visualize the epidemiologic movement of case, death, and vaccination rates through maps, tables, and charts.

Credit: Johns Hopkins University Center for Systems Science and Engineering (CSSE). Data source: Johns Hopkins University, Esri, FAO, NOAA, USGS.

CHAPTER 20

Sharing your work

Okay, are you ready for an easy chapter about a cool topic? ArcGIS Pro makes it easy to share your work. You can share within your own organization, with colleagues elsewhere, or with the public. You can share just the plain data or share layers with all their symbolization, labeling, and so on.

If you symbolize a layer and want to share the symbology with someone who has the same source data (for example, if you are both accessing the same data from a shared location), you can share a layer file (.lyrx). Layer files are super handy to use. For example, suppose you had a zoning layer with 15 zoning categories symbolized in a certain way. It would take a lot of work to re-create that symbology every time you add that layer to a new project. Applying symbology from a layer file can help keep your work consistent. When you start a new project, you can add the layer file to a map, just as you would a feature class. The beauty of this file is that it's added to your new project already symbolized. Remember, this file contains only the look of your data; if you send this file to someone, that person will need the data, as well. You can package these elements together in a handy file called a layer package (.lpkx), which includes both the layer properties and the dataset referenced by the layer.

Chapter 20: Sharing your work

You can share a map by packaging your data with the map information in a map package (.mpkx). A map package contains a map and the data referenced by its layers, so it can be easily shared. You can also use map packages to create a backup of a particular map with a snapshot of the current data used in the map.

You can even share entire projects by packaging all your project data. This package is called a project package (.ppkx) and includes the project file (.aprx) and all your data, layers, maps, and layouts. Magic! Well, it seems magical, but it's also just cool technological wizardry. Either way, we like it!

Hope this sounds good, because you're going to try all of them.

Export a layout

You're going to open an earlier project and share from there.

1. Start ArcGIS Pro, and click Open Another Project. Browse to C:\GIS20\Chapter05\Ch5_thematic, and click the Ch5_thematic.aprx project. (You may get lucky and find it in your list of recent projects.)

   An easy way to share a map is to export your finely crafted layout to a .pdf file. This file is a common format and simple to print.

2. If needed, click the Layout tab to activate it.

3. On the ribbon, on the Share tab, in the Output group, click Export Layout to open the Export Layout pane.

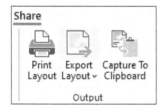

4. For File Type, click PDF.

5. For Name, browse to C:\GIS20\Chapter20, and save with the name **MyLayout**.

   Plenty of other options are available in the Export Layout pane, but we're going to keep it simple and keep the defaults.

6. Click Export at the bottom of the pane.

   You'll know it's finished when the Export Completed message appears at the bottom of the pane.

   Let's see what the file looks like.

7. Click View Exported File.

   Sure enough, your layout looks good!

Create a layer file

You did some good work symbolizing and labeling the CountiesAge layer. Now a colleague wants to use the same symbology. What a compliment! They have access to the data in a shared location, so all you need to send them is the .lyrx file.

1. Click the Map tab so the map is active.

   It's always best to save the layer file (.lyrx) to the same folder as the underlying data. In this case, you're going to create a layer file that references CountiesAge, so you need to know the location of that file in your folder structure.

2. At the top of the Contents pane, click the List Data by Source button to locate the folder containing CountiesAge.

3. In the Contents pane, right-click CountiesAge, and click Sharing > Save as Layer File.

Chapter 20: Sharing your work

4. Browse to C:\GIS20\Chapter04.

   A good practice is to keep the name similar to the original shapefile but add something to remind you what the file is symbolizing.

5. Name the layer file **CountiesAge_Seniors**, and click Save.

Add the layer file to your map

Not much seems to have happened, but you did save that file. Let's see whether it worked.

1. On the Insert tab, in the Project group, click New Map to add a new map.

2. Click Add Data, browse to where you saved the layer file, and add the CountiesAge_Seniors.lyrx file.

   Look at that! There's your perfectly symbolized and labeled layer in the new map, Map1. Pretty cool! As long as the data source is available to your colleague, it's that easy to share your symbology.

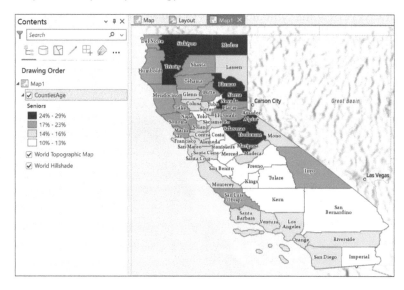

Import symbology from a layer file

Sometimes you encounter a layer file that doesn't open correctly because it's disconnected from its original data. Fear not! You can still apply its properties.

1. In the Map1 Contents pane, right-click CountiesAge, and click Remove to remove it from the map.

2. Click Add Data, browse to C:\GIS20\Chapter04, and add CountiesAge.shp.

   Because you're adding the plain, unsymbolized shapefile, it draws on the map without the symbology you specified.

3. In the Contents pane, right-click CountiesAge, and click Symbology.

4. In the Symbology pane, click the options button (three stacked lines) in the upper right, and click Import Symbology.

5. In the Apply Symbology from Layer pane, for Symbology Layer, browse to C:\GIS20\Chapter04, and click the CountiesAge_Seniors.lyrx file.

   You should always review the symbology fields to make sure that the correct field (Seniors, in this case) will be symbolized.

Chapter 20: Sharing your work

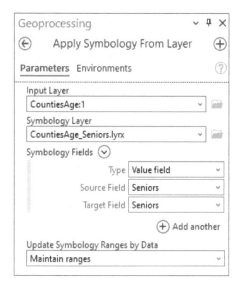

6. Run the tool.

    Voilà! The Seniors symbology and labeling have been properly applied to the layer. This capability would be a lifesaver in a 15-category zoning layer or a series of maps with the same symbology.

Create a map package

Suppose a colleague at another organization wants to use your ArcGIS Pro map. Another compliment! Sure, you can totally do that.

1. On the ribbon, on the Share tab, in the Package group, click Map to open the Package Map pane.

2. In the Package Map pane, under Start Packaging, click Save Package to File.

    Good to know that you can save it to ArcGIS Online, too.

3. For Name, browse to C:\GIS20\Chapter20, and name the map package **Seniors**. Click Save.

4. Optionally, add summary information and tags used for searching.

5. Under Options, check the Include Enterprise and UNC Path Data box.

   Including enterprise and UNC path data is a good way to be sure all your data is included in the map package.

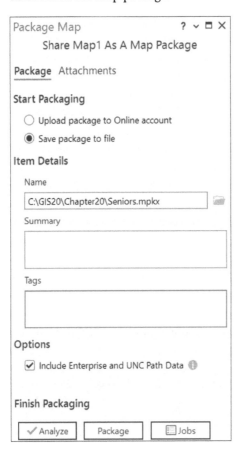

6. Under Finish Packaging, click Analyze to verify that your specifications are valid. Your results should say, "No errors or warnings found."

7. Click Package to create your map package.

   Creating the map package takes several moments to finish.

8. When you receive the success message, save and close the project.

Open the map package

You got a success message, so we feel good about that, but let's check to make sure it worked.

1. In File Explorer, browse to C:\GIS20\Chapter20, and double-click the Seniors.mpkx file.

   Look, it even has a little icon indicating that it's a map package file.

   Opening the map package takes a moment, but when it opens, you'll notice a few things. The package contains only Map1, not the other map frame or the layout.

2. Close this untitled project without saving.

Create a project package

Now suppose a colleague at another organization wants to use your entire ArcGIS Pro project. You guessed it—you can do that, too.

1. In ArcGIS Pro, open the Ch5_thematic project.

   > You should be able to find it in the list of recent projects this time.

2. On the ribbon, on the Share tab, in the Package group, click Project to open the Package Project pane.

3. In the Package Project pane, for Start Packaging, click Save Package to File.

4. For Name, browse to C:\GIS20\Chapter20, and name the project package **Seniors**.

5. For Summary, type **Project of percentage of seniors in California**. For Tags, type **seniors** and **California**.

6. Check the Share Outside of Organization box. Confirm that Include Toolboxes and Include History Items are both checked.

7. Click Analyze to verify that your specifications are valid. Your results should say, "No errors or warnings found." Fix any errors, if needed.

8. Click Package to create your project package.

   Like a map package, creating a project package takes several moments to finish.

9. When you receive the success message, save and close the project.

Open the project package

Again, let's check to be sure it worked.

1. In File Explorer, browse to C:\GIS20\Chapter20, and double-click the Seniors.ppkx file to open it.

   The project package contains both maps and the layout. Now you can share your entire project with your colleague.

2. Save and close the project.

   That was fun! You learned how to share a layer (or make life easier for yourself), a map package, and an entire project. That's good stuff for today. Now get back to your other work!

USER STORY

Using GIS to map bikeshare and e-scooter systems

In this chapter, you learned about sharing data, maps, and projects on a small scale. That's certainly all you need to know at this stage. But now you'll read about how a large federal government agency approached sharing GIS on a grander scale. Aren't you glad you don't have to deal with those struggles?

The US Department of Transportation (DOT) had a significant opportunity to better structure its GIS operations. Because no centralized architecture existed, operational and program offices found their own solutions to storing and processing geospatial data. Using GIS, DOT teams have helped meet goals and provide what can't be discerned with tabular data, providing aha moments for management and stakeholders.

Your authors think DOT still puts out some great work. Check out the *Interactive Bikeshare and E-Scooter Map* at links.esri.com/BikeScooter.

This interactive bikeshare and electric scooter map provides the name of the docked or dockless bikeshare and e-scooter system in major US cities from 2015 to the present. In July 2022, 61 docked bikeshare systems and 8,473 docking stations were open to the public.

Source: US Department of Transportation, Bureau of Transportation Statistics. Data sources: Mapbox, OpenStreetMap, Office of Spatial Analysis and Visualization at the Bureau of Transportation Statistics.

CHAPTER 21

Publishing your work (bonus skill)

Because ArcGIS Pro is so tightly integrated with ArcGIS Online, you can also share your work by publishing it to the web. An ArcGIS Pro data layer can be published as a web layer, which can then be used in a web map in ArcGIS Online. You can also publish a map made in ArcGIS Pro as a web map to ArcGIS Online. From there, you can publish your web map to ArcGIS StoryMaps℠, a dashboard, or even ArcGIS Experience Builder.

Publishing your map is for more than just your fellow GIS enthusiasts. You can share with your manager information about properties your company is considering buying. You can share an analysis of homelessness with policy makers. You can show your friends the trip you took to Rome in the form of a StoryMaps story.

With ArcGIS StoryMaps, there are so many interesting stories that you can easily get lost in them and spend more time than you intended. We certainly have! If you have time to spare, check out the ArcGIS StoryMaps Gallery at links.esri.com/StoryMapExamples.

Let's take a step back to show you how to publish your layers and maps.

Share a web layer

The friends in your hiking club heard about your trailheads layer and want you to publish it for them to see.

1. In ArcGIS Pro, click Open Another Project. Browse to C:\GIS20\Chapter10, and open the Ch10_trailheads project.

2. In the Contents pane, click the Trailheads_XYTableToPoint layer to activate it.

3. On the ribbon, on the Share tab, in the Share As group, click Web Layer.

   The Share as Web Layer pane opens.

4. In the Share as Web Layer pane, for Name, type **Trailheads**.

5. For Summary, type **Trailheads in the Boulder, Colorado area**.

6. For Tags, type **trailheads** and **Colorado**.

   > **Hint:** Add a comma or press the Tab key after each tag. If you add just a space, just one tag of "trailheads Colorado" will be created instead of two separate tags.

7. For Layer Type, keep Feature selected.

8. For Location, keep Folder empty.

9. For Share With, check the box for Everyone.

   The Everyone option shares the web layer with your organization, as well. Your organization's name should appear instead of "Organization" in this pane.

Chapter 21: Publishing your work (bonus skill)

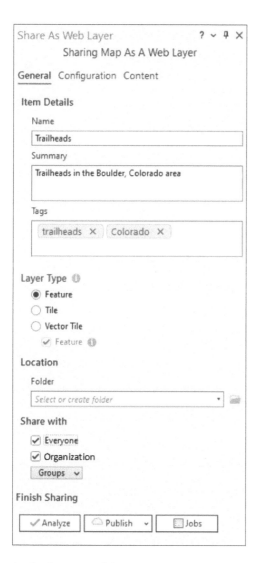

10. At the bottom of the pane, click Analyze.

The analysis generated one error and two warnings. Items with the white *X* in the red circle are errors, which must be resolved to continue. An exclamation point in a yellow triangle is a warning, which you don't need to resolve but may choose to. A web layer can't be published with an error, so let's fix it.

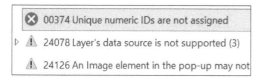

11. In the Share as Web Layer pane, click the three dots to the right of the error text to see your options for fixing the error.

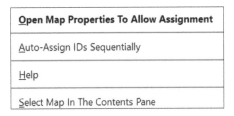

12. Click Open Map Properties to Allow Assignment.

    The Map Properties dialog box opens.

13. In the Map Properties dialog box, on the General tab, check the Allow Assignment of Unique Numeric IDs for Sharing Web Layers box.

    A message appears at the bottom, but that's okay.

14. Click OK to save the new map property.

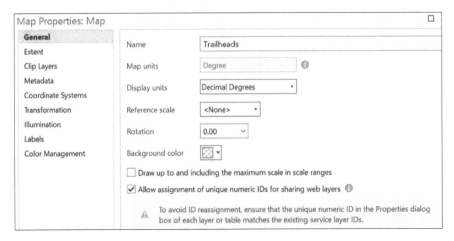

The Share as Web Layer pane now shows a green check mark, indicating that the error has been resolved.

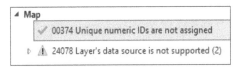

You're going to ignore the warning messages because they don't prevent you from publishing the layer.

15. Click Publish.

    When publishing is complete, a message saying that the web layer published successfully appears in the pane.

View your web layer contents in ArcGIS Online

You'll sign in to your ArcGIS Online account to confirm that your web layer was published.

1. In a web browser, go to **arcgis.com**.

2. Sign in with your username and password.

3. Click the Content tab at the top.

    Under My Content, you should see your Trailheads web layer, called a feature layer (hosted) here, at the top of your contents list. It's shared with everyone (globe icon) just as you specified.

    A service definition file was also created. The service definition contains all the information needed to publish a web layer or service, such as properties, capabilities, and service type. It's a handy reference that's generated by ArcGIS Pro automatically. The service definition takes on the sharing properties of the feature layer, so it's also shared with everyone.

4. Click the Trailheads feature layer to open the item page.

    On the item page, the summary and tags appear as you typed them in the Share as Web Layer pane. A thumbnail is also provided.

View your web layer in Map Viewer

1. In the upper right, click Open in Map Viewer.

   The Trailheads layer opens in a new web map.

   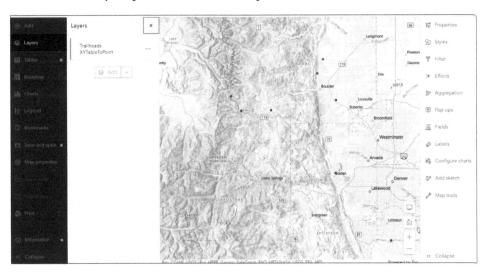

   We aren't going to cover the subject of Web GIS in this book (that could be, and in fact is, another book—*Getting to Know Web GIS*, fifth edition, by Pinde Fu [Esri Press, 2022]), but we'll tour Map Viewer. Map Viewer is similar to the layout in the ArcGIS Pro interface: the map appears in the middle, the Contents (dark) toolbar appears to the left, and the Settings (light) toolbar appears to the right. Under the Trailheads layer is the Add button. This button is the web version of your good buddy in ArcGIS Pro, the Add Data button. In the Layers pane, the Trailheads layer has an Options button (three dots) next to it.

2. Click the Options button for Trailheads, and examine the options.

   The options include some of the same options that appear when you right-click a layer in the ArcGIS Pro Contents pane.

3. To the left, examine the options on the Contents toolbar.

   These familiar items include Layers, Tables, Basemaps, Charts, Legend, Bookmarks, Save and Open, and Map Properties.

4. To the right, examine the options on the Settings toolbar.

   The settings should be familiar to you, as well—Properties, Styles, Pop-ups, Labels, and so on. Your ArcGIS Pro skills should translate fairly easily, allowing you to do some good work in ArcGIS Online, too.

5. On the Contents toolbar, click Save and Open > Save As.

6. For Title, type **Trailheads**, and click Save.

   Next, we're going to show you how to share your ArcGIS Pro map as a web map, so let's get out of this web map now.

7. Click the menu button (three stacked lines) to the left of Trailheads, and click Content.

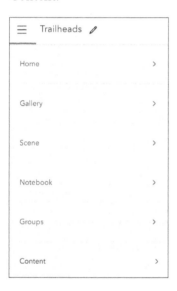

   Clicking Content takes you back to the ArcGIS Online My Content tab.

8. Minimize the browser window for now.

Share a web map

1. Return to ArcGIS Pro.

    You're going to share an entire web map, which includes the trailheads and the stream gauges.

2. On the ribbon, on the Share tab, in the Share As group, click Web Map.

    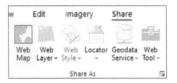

    The Share as Web Map pane appears. It looks a lot like the Share as Web Layer pane.

3. For Name, type **Trailheads and Stream Gauges**.

4. For Summary, type **Trailheads and streams in the Boulder, Colorado area**.

5. For Tags, type **trailheads**, **stream gauges**, and **Colorado**.

6. For Select a Configuration, click the down arrow to examine the choices, but keep the default set to Copy All Data: Exploratory.

7. For Location, keep the Folder text box empty.

    This parameter lets you organize your ArcGIS Online content into folders, which is helpful if you have a lot of content but is not something you need to worry about yet.

8. For Share With, check the Everyone box.

    The name of your organization will appear on the line below that.

Chapter 21: Publishing your work (bonus skill)

9. At the bottom of the pane, click Analyze.

   This time, two errors and one warning are generated. Let's fix them.

10. Click the error down arrow to read the errors.

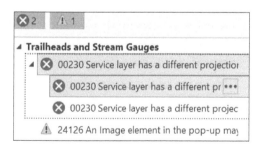

Fix errors

1. On the first error, click the three dots to the right of the error text.

2. Click Update Map to Use Basemap's Coordinate System.

Update Map To Use Basemap's Coordinate System
Change Coordinate Systems Properties
Help
Select Layer In The Contents Pane

   The second-level error now shows a green check mark. Yay, that was easy to fix! You can return to that menu and click Help to read about why that error happened.

3. Repeat these two steps with the remaining second-level error.

   A second green check mark appears. Now all the error messages are resolved.

4. At the bottom of the pane, click Share.

   When your web map has successfully uploaded, ArcGIS Pro generates a success message at the bottom of the pane.

View your web map in ArcGIS Online

You'll return to ArcGIS Online to confirm that your web map was published.

1. Maximize the browser window to see ArcGIS Online.

   Weird. The new web map isn't in My Content.

2. Refresh your browser window.

   The Trailheads and Stream Gauges web map now appears in My Content at the top of the list. It's shared with everyone, just as you specified.

☐ Title		
☐ Trailheads and Stream Gauges	🖼 Web Map	◐
☐ Trailheads and Stream Gauges_WFL1	📍 Feature layer (hosted) ▼	◐
☐ Trailheads and Stream Gauges_WFL1	📄 Service definition	🔒
☐ Trailheads	🖼 Web Map	🔒
☐ Trailheads	📍 Feature layer (hosted) ▼	◐
☐ Trailheads	📄 Service definition	🔒

View your web map contents in ArcGIS Online

1. Click Trailheads and Stream Gauges next to Web Map to open the item page.

   The summary and tags you entered appear, as do the Trailheads_XYTableToPoint and StreamGaugesLive layers and Trailheads.csv under Tables. You may notice that the word *Authoritative* appears beside the StreamGaugesLive layer. Remember how you originally added that layer from ArcGIS Living Atlas to your chapter 10 map in ArcGIS Pro? This layer streams (no pun intended) live data from ArcGIS Living Atlas, so when stream conditions change, this layer updates in your web map—and lets your hiking club know the current conditions.

   The web map just has a generic thumbnail, but you can easily customize it.

2. Click Edit Thumbnail above the generic thumbnail of the United States.

3. In the Create Thumbnail dialog box, click Create Thumbnail from Map.

4. When the dialog box changes to a map view, pan or zoom as you want.

5. Click Save to save your web map's thumbnail.

   Now your Trailheads and Stream Gauges web map has its own custom thumbnail.

View your web map in Map Viewer

1. On the Trailheads and Stream Gauges item page, click Open in Map Viewer in the upper right.

   The map opens in Map Viewer.

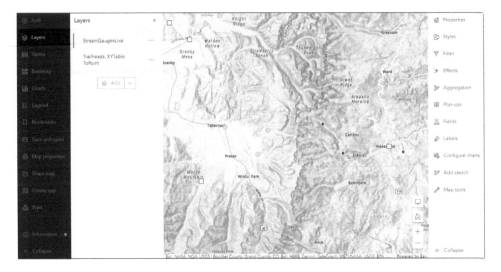

From here, you can go crazy with changing symbology, adding labels, and configuring pop-ups. When the web map is ready, you can share it in an app.

Explore options for creating an app

1. On the Contents toolbar in ArcGIS Online, click Save and Open > Save to save your web map.

2. On the Contents toolbar, click Create App.

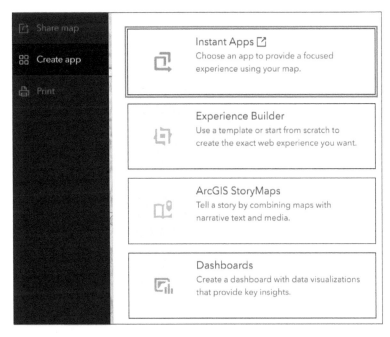

This is where the fun really begins, but it's also an entirely new topic! For now, you've learned that publishing layers and maps to ArcGIS Online can let you take your ArcGIS Pro mapping skills to another level. Until you acquire those skills, you can copy the URL from your browser and send this map to your hiking club.

3. Save and close ArcGIS Pro.

Conclusion

Congratulations! You've finished the book and learned so many essential GIS skills. Sadly, that brings us to the end of our journey together! We hope this book can be an ongoing reference for you as you move forward with GIS.

Now you can put on your résumé that you've worked with ArcGIS Pro and ArcGIS Online and have a basic understanding of GIS. You know how to get data that's GIS ready, import data that has addresses or latitude and longitude, enrich that data from a vast well of information, analyze the data based on attributes or spatial relationships, symbolize your results, and share them with your group or the world. That's impressive!

Over the years, our students have used these skills to address a variety of issues—from serious issues such as where to locate a clinic for people with asthma to lighter issues such as where people should live after graduate school. At this point, if you're interested in reinforcing your GIS skills, check out the gallery of free tutorials at learn.arcgis.com. You'll find tutorials ranging from introductory to advanced on a variety of topics. You have covered no small amount of ground already, and you have undoubtedly recognized that the field of GIS is widespread, valuable, and—dare we say it?—fun!

USER STORY

Mapping climate to protect people and the environment

Climate Mapping for Resilience and Adaptation (CMRA) integrates information from across the federal government to help people consider their local exposure to climate-related hazards. People working in community organizations or for local, tribal, state, or federal governments can use the site to help develop equitable climate resilience plans to protect people, property, and infrastructure. The site also points users to federal grant funds for climate resilience projects, including those available through the Bipartisan Infrastructure Law.

The CMRA website (resilience.climate.gov) is a model for publishing GIS content. It provides web maps with accompanying real-time statistical data about where people, property, and infrastructure may be exposed to hazards. It features an assessment tool that helps users understand exposure in their area so they can plan and build more resilient infrastructure in their community. The site also provides data and tools to help users with climate resiliency planning.

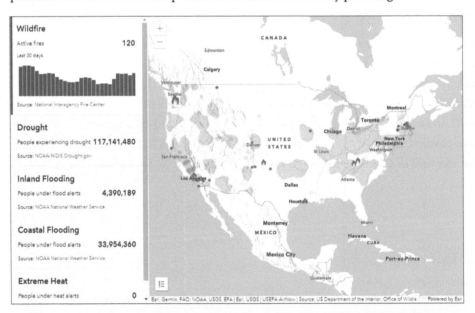

Web app technology on the CMRA site is valuable for visualizing climate-related hazards, such as wildfires and drought, on continually updated web maps. With the CMRA assessment tool, users can learn about how climate may change in their area.

Credit: CMRA. Data sources: CMRA, Esri, Garmin, FAO, NOAA, USGS, EPA, USEPA AirNow, US Department of the Interior, Office of Wildland Fire, NIDIS, Drought.gov, NWS.

ABOUT ESRI PRESS

Esri Press is an American book publisher and part of Esri, the global leader in geographic information system (GIS) software, location intelligence, and mapping. Since 1969, Esri has supported customers with geographic science and geospatial analytics, what we call The Science of Where®. We take a geographic approach to problem-solving, brought to life by modern GIS technology, and are committed to using science and technology to build a sustainable world.

At Esri Press, our mission is to inform, inspire, and teach professionals, students, educators, and the public about GIS by developing print and digital publications. Our goal is to increase the adoption of ArcGIS and to support the vision and brand of Esri. We strive to be the leader in publishing great GIS books, and we are dedicated to improving the work and lives of our global community of users, authors, and colleagues.

Acquisitions
Stacy Krieg
Claudia Naber
Alycia Tornetta
Craig Carpenter
Jenefer Shute

Editorial
Carolyn Schatz
Mark Henry
David Oberman

Production
Monica McGregor
Victoria Roberts

Sales & Marketing
Eric Kettunen
Sasha Gallardo
Beth Bauler

Contributors
Christian Harder
Matt Artz
Keith Mann

Business
Catherine Ortiz
Jon Carter
Jason Childs

Related titles

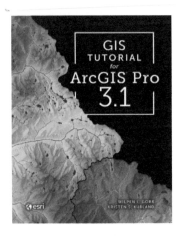

GIS Tutorial for ArcGIS Pro 3.1
Wilpen L. Gorr and Kristen S. Kurland
9781589487390

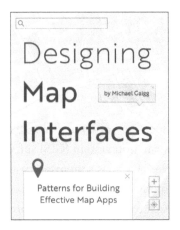

Designing Map Interfaces: Patterns for Building Effective Map Apps
Michael Gaigg
9781589487253

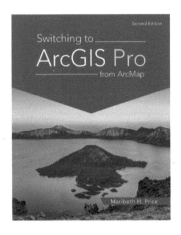

Switching to ArcGIS Pro from ArcMap, second edition
Maribeth H. Price
9781589487314

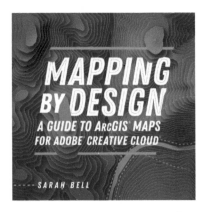

Mapping by Design: A Guide to ArcGIS Maps for Adobe Creative Cloud
Sarah Bell
9781589486041

For information on Esri Press books, ebooks, and resources, visit our website at
esripress.com.